RISK BY CHOICE

RISK BY CHOICE

Regulating Health and Safety in the Workplace

W. KIP VISCUSI

HARVARD UNIVERSITY PRESS
CAMBRIDGE, MASSACHUSETTS
AND LONDON, ENGLAND
1983

LIBRARY OF CONGRESS CATALOGING IN PUBLICATION DATA

Viscusi, W. Kip.
Risk by choice.

Bibliography: p.
Includes index.
1. Industrial hygiene—United States.
2. Industrial safety—United States.
3. Industry and state—United States.
I. Title.
HD7654.V57 1983   363.1′15′0973   82-15591
ISBN 0-674-77302-0

PREFACE

GOVERNMENT REGULATION OF HEALTH AND SAFETY RISKS HAS been characterized by a myopic concern with risk reduction, irrespective of the magnitude of the risk or the cost of eliminating it. Although there is an emerging consensus that these policies should be more balanced, the narrowly defined legislative mandates of the risk regulation agencies have impeded efforts to broaden their focus. My purpose in this volume is to indicate the proper role of regulation in the market context where these policies must operate.

As the title of this book suggests, I have structured my approach in terms of a quite definite point of view: that many risks are the outcome of individual choices, such as the worker's decision to accept a potentially hazardous job. The basis for government intervention to alter the results of these decisions derives largely from inadequacies in the way workers make these choices, particularly because of their lack of knowledge concerning the risks they face. Although there is a potentially productive role for government policy, market processes can generate powerful incentives for safety. We should design policies to complement these economic forces rather than supplant them. Moreover, the structure of regulatory policies should take into account the cost of the regulations and the value workers would place on the reduction in risk if they were fully cognizant of the risks they faced.

These themes have been the foundation of my more technical

work, and I have attempted to promote them in my policy-related activities. From 1979 to 1980 I served as the deputy director of the President's Council on Wage and Price Stability, which was responsible for the White House's regulatory oversight activities during the Ford and Carter administrations. My work for the council sharpened my assessment of which economic principles are most fundamental to the efficient design of risk regulations and enabled me to make a more informed evaluation of the merits of the regulatory oversight effort, which is the subject of Chapter 8.

Although I have attempted to bring the policy-related material up to date, it is still too soon to make precise judgments about policies instituted by the Reagan administration. The problem arises in part because recent statistics, for example on accident rates and OSHA operations, are not yet available. Nevertheless, one can make some judgments regarding the effectiveness of the new structure for White House oversight of regulations and the implications of the shifts in risk regulation policies initiated by regulatory agencies.

During my eight years of research, I published many of my research findings in my book *Employment Hazards* and in articles in technical journals. Because this work is accessible mainly to the small group of economists working in this area, I have synthesized my results to reach a much broader audience, including policymakers and others who are concerned with the impact and design of risk regulation policies. This book also includes a considerable amount of my more policy-oriented work, much of which has not been published previously. Although I have attempted to keep the technical demands on the reader relatively modest (for example, no equations appear in the text), I do assume that the reader is familiar with risk regulations and that he is willing to think systematically about the design of risk regulation policies.

Since this book is the outcome of a long period of research, I will not attempt to provide an exhaustive list of the people who have helped me. As IBM Research Professor at the Fuqua School of Business at Duke University, I was freed from much of my teaching obligation during 1981–82 and thus had the time to complete this work. The timely preparation of the manuscript was made possible by the superb typing of Virginia Bostrom at Northwestern University and of Kathi Sturtevant at Duke, who coped with endless revisions. My greatest immediate and long-term debts are to members of my family, whose general support I would like to acknowledge here.

Some portions of this book appeared previously in several articles. I would like to thank the editors of the following publications for permission to draw on this work: *Public Policy, Journal of Occupational Medicine, Journal of Policy Analysis and Management* (Greenwich, Conn.: JAI Press, 1981), and *Research in Public Policy Analysis and Management.*

CONTENTS

RISK BY CHOICE

1

RISK AND INDIVIDUAL CHOICE

WHOSE LIFE IS WORTH HOW MUCH? TO SOME IT SEEMS IMMORAL even to ask, but to others—to the worker, say, who is offered a dangerous but lucrative job—the question inevitably presents itself. Framers of public policy who deal with this problem often seem to be playing God. Should they interfere with a worker's decision, his own negotiaton with destiny? If they interfere, when should they do so and how should they do it?

This is not a simple problem. Some risks are incurred knowingly. Other risks, such as those we inherit, are not subject to individual discretion. Risks in the workplace are seldom at either extreme. Workers often have at least partial knowledge of the hazards they face, but in some instances they may encounter an unanticipated rendezvous with destiny. These market-traded risks are the result of individual choices, and in their regulation individual choice plays a fundamental role.

Although the recent flurry of interest in risk regulation might suggest that these issues have arisen only recently, they are not new. What has changed most dramatically is the diversity of the risks we face and our understanding of their nature. The government's response to these risks has also escalated, with many new regulatory agencies charged with the task of reducing risks.

These regulatory interventions share many common elements. Typically, a regulatory agency identifies a hazard, declares that it is harmful and should be eliminated, then establishes standards to mandate the reduction of the risk to the lowest possible level. The strategy for job risk regulation followed by the Occupational Safety and Health Administration (OSHA), which is the focus of much of this volume, epitomizes this approach.

My decision to treat job hazards as a generalizable paradigm for risk regulation follows a reasonably well-established economic tradition. Since the time of Adam Smith, employment hazards have served as the standard example for analyzing how markets respond to perceived risks. The role of decisions by workers and firms in generating observed levels of safety parallels the function of market forces in other situations in which risks are the consequence of economic choices. A major reason for not focusing on some other type of risk is that we have extensive data on pertinent aspects of worker behavior, enabling us to quantify the role of these economic influences. Indeed, almost all recent studies of the appropriate value society should place on life have focused on the labor market, since we have comprehensive information on employment patterns in hazardous occupations.

Also, job risks have been a principal target for government regulation. OSHA's approach to risk regulation is representative of that of many other agencies, in part because OSHA's prominence has given it a leading policy role. Chapters 2 and 9 are directed at specific aspects of OSHA policies as well as the broader features of this regulatory approach. In other chapters I deal much more generally with the fundamental principles for designing risk regulation, using employment hazards to illustrate these principles. Rather than catalogue all possible parallels with other types of risks through a series of case studies from different agencies, I have chosen a more methodological orientation in which I identify the salient principles involved.

From its inception, the federal government's effort to regulate job risks and other hazards has been fundamentally ill-conceived, in large part because these regulations have not been based on principles for efficient policies. Congress created OSHA to abate the rapid upsurge in recorded job accidents in the 1960s—an increase that, as I will demonstrate, was largely a statistical artifact; job hazards actually were continuing to diminish throughout this period, as were other risk levels. OSHA's legislative directive to reduce job risks to the lowest technically feasible levels, and the implementation of this mandate, assumed an absolutist character. All jobs with nonzero risk levels became candidates for regulation.

This and similarly inspired regulatory efforts have had a deleterious economic effect because they are based on the rather simplistic view that risks are generated by mistaken technological choices, which are believed to be entirely independent of cost considerations. Somewhat surprisingly, the same market critics who display a well-founded skepticism of the efficacy of new technologies almost invariably seek technological solutions to risk problems. This attitude is reflected in OSHA's penchant for engineering controls and its emphasis on policies requiring that workplace technologies reduce risks to their lowest feasible or lowest measurable level. Indeed, the unbounded faith in technology is reflected in the common practice of completely ignoring the cost of the mandated changes.

The neglect of the costs of risk regulations also stems from a failure to acknowledge that there are important tradeoffs involved. Risk regulation advocates have consistently claimed that lives should not be bartered for dollars. Such an uncompromising assertion may make effective political rhetoric, but policies based on this absolutist approach have produced no perceptible health benefits to workers and have inflicted a significant burden on society.

The central difficulty with the traditional approach to risk regulation is the assumption that a no-risk society is desirable. The policy approach I advocate here attempts to promote efficient, nonzero risk levels by augmenting existing market forces. If workers understand the risks they face on hazardous jobs, they will demand additional compensation to take such jobs. My analysis of the wage-risk tradeoffs reflected in decisions by workers and firms suggests that these choices produce powerful incentives for safety—on the order of almost $70 billion annually. This figure represents only the wage compensation workers receive for risk and does not include investments in safety stimulated by these market forces. These incentives are bolstered by the effects of workers who quit in response to hazards they have learned about after taking the job. Since existing market forces generate financial incentive for risk reduction that dwarf the incentives created by OSHA, one should be careful not to displace these productive aspects of market behavior in our zeal to control the risks that remain.

Notwithstanding these beneficial aspects of market behavior, one can advocate a potential role for government risk regulation, largely because workers and firms are not aware of the implications of all risks that are present in the workplace. When workers are not cognizant of the risks they face, their employment decisions no longer represent an implicit choice to accept them. Nevertheless, even imperfect market forces create powerful incentives for safety. Government policies directed at alleviating market deficiencies can be most effective if they take advantage of existing market mechanisms rather than trying to supersede these forces. Providing job risk information to workers to enable them to select jobs that are appropriate for their own risk preferences is a promising strategy for augmenting market forces and promoting outcomes that are responsive to the diversity of attitudes toward risk.

Individual choices also provide guidelines for analyzing the impact and desirability of risk regulation policies. The link between policy design and policy outcomes is governed by the responses of workers and firms to the incentives created by the policy. Once the impacts are known, we still need some criterion for assessing the merits of different policies, particularly what risk–money tradeoff we should adopt. A useful starting point for this assessment is the tradeoff displayed in the choices of the workers themselves.

To assess accurately the shortcomings of present risk regulation policies, we must also explore the political context in which they have developed, which is my primary concern in the latter part of the book. The failure of the risk regulation agencies and the need for a major redirection of their efforts inevitably raise the question of why these misguided policies have not been brought under control. The need for doing so was the impetus behind President Ford's establishment of the White House oversight process. As my review of the regulatory oversight mechanisms will indicate, these efforts have been hampered by insufficient political authority and by the legislative mandates of the regulatory agencies. The inadequacies of the oversight process must be remedied before risk regulation policies can be reformed, since oversight is the principal institutional mechanism for stimulating regulatory reform. To be successful, proposed changes in regulations must be coupled with a change in the political context in which regulators operate.

By fostering market control of job risks rather than attempting to operate independently of market forces, the approach to risk regulation that I advocate will promote levels of safety that better reflect the risk levels workers would select if they were completely aware of the consequences of their choices. Although policymakers will no longer be able to make unqualified commitments to risk reduction, the more balanced policies that emerge will do more to promote worker welfare than have those based on the false hope of a no-risk society.

2

OSHA: Design, Implementation, and Impact

AFTER A SOCIAL PROBLEM IS IDENTIFIED, THE USUAL GOVERN-
ment response is to mount a direct attack on it. We provide income support to the poor, food stamps to the hungry, and medical care to the ill. Similarly, the problems associated with
occupational hazards have led the government to set standards to
ensure minimal levels of safety.

OSHA's Legislative Mandate

OSHA's approach to regulation was largely defined by its enabling legislation, the Occupational Safety and Health Act of
1970. The objective of the agency, which began operation on
April 28, 1971, was "to assure so far as possible every working
man and woman in the Nation safe and healthful working conditions."[1] The policy tool for promoting this objective was "to set
mandatory occupational health and safety standards applicable
to business affecting interstate commerce."[2] If these standards
were not exhaustive, employers were subjected to an additional,
more general requirement to "furnish to each of his employees
employment and a place of employment which are free from rec-
ognized hazards that are causing or are likely to cause death or
serious physical harm to his employees."[3] Much of OSHA's impact has been determined by the overall strategy for intervention—the reliance on standards, the levels of stringency for these
standards, and the types of standards utilized.

Implementation of this legislation has reinforced its strongly stated directives. Agency officials have subscribed to a narrow interpretation, arguing that a standard should be tightened to the lowest feasible level of risk so long as there is any possibly beneficial effect of increased stringency. The feasibility considerations that are permitted to enter have been restricted to technical feasibility and affordability to the firm rather than to a broader economic feasibility concept that assesses the costs and benefits of a proposal.

Throughout almost the entire first decade of OSHA's operations, critics have argued for a more balanced interpretation of OSHA's mandate. Since the enabling act included qualifiers such as ensuring safety "as far as possible," providing medical criteria to promote worker health "insofar as practicable," and setting "feasible" health standards, OSHA may have sufficient leeway to formulate policies based on their costs and the extent of benefits rather than solely on the existence of some beneficial health impact, however small.[4] The agency could have taken the initiative by proposing changes in the legislation or by advocating a liberal interpretation of its mandate, leaving it to the courts to impose a more stringent basis for regulation.

Instead, agency critics have had to force OSHA to broaden its concerns. One source of pressure has been the White House regulatory oversight process, which will be discussed in the last two chapters. Since November 1974, regulatory agencies such as OSHA have been required to calculate the costs and economic effects of major new regulations. Although this process has prevented some of the most ill-conceived regulations from being put into effect, the regulatory review process has not yet been vested with sufficient authority to alter the overall thrust of OSHA policy.

Efforts to have the courts force OSHA to make balanced economic judgments have failed to remedy its deficiencies. The two principal U.S. Supreme Court decisions have been in the benzene and cotton dust cases, neither of which fully resolved the nature of OSHA's obligations.[5] The 1980 benzene decision sidestepped the benefit-cost tradeoff issue, focusing instead on whether OSHA had demonstrated that the standard would generate significant reductions in risk. In particular, the Supreme Court af-

firmed the Court of Appeals judgment that the benzene standard
had not been shown to be "reasonably necessary or appropriate
to provide safe and healthful employment." Moreover, the Court
rejected the argument that OSHA's health and safety mandate
was absolute: "But 'safe' is not the equivalent of 'risk-free.' A
workplace can hardly be considered 'unsafe' unless it threatens
the worker with significant risk or harm."[6]

Unfortunately, the Court did not go beyond simply maintain-
ing that OSHA must focus on "significant" risks until the 1981
cotton dust decision, when it specifically rejected the use of a
benefit-cost test for regulations pertaining to toxic materials or
harmful physical agents. The Supreme Court upheld the cotton
dust standard by interpreting the feasibility provisions of
OSHA's enabling legislation in terms of the technical possibility
of compliance ("capable of being done") rather than in terms of
broader economic feasibility.

It is difficult to reconcile the implications of these decisions
with any meaningful principles for policy design. The signifi-
cant-risk concept endorsed by the Court presumably includes
consideration of the probability, as well as the severity, of the
adverse outcome. A conventional benefits analysis simply imposes a severity-weighting scheme based on the value of these
risks to society. By weighting risks in this fashion, one can evaluate policies using a cost-effectiveness framework that may yield
policy rankings identical to those in a formal benefit-cost analysis—an approach which the Court has since rejected.

How far OSHA can go in considering costs is unclear, particularly since technical feasibility is closely linked to the economic
cost of acquiring existing technologies. More broadly, firms must
select the amount of money they will invest in developing new
technologies before deciding whether it is preferable to violate
the standard or cease operations. Is compliance with a standard
feasible if some finite level of expenditure would bring the firm
into compliance? If these expenditures are so high that they
would lead to the closure of some or all firms in the industry, is
the standard nevertheless feasible? Quite simply, there is no
meaningful way to assess technical feasibility apart from the associated costs.

The Court's conclusion that Congress has structured OSHA's mandate to prevent explicit cost-risk tradeoffs for certain classes of risk may have been an accurate interpretation, but this narrow approach did not serve as the basis for the cotton dust standard, which the Court upheld. The level of the standard was broadly consistent with a cost-effectiveness framework in which cost-risk tradeoffs guided the standard-setting process (see Chapter 7), so standards were set above their lowest technically feasible risk levels. Moreover, OSHA designs and enforces its standards in a manner that will prevent the closure of firms, so cost considerations do affect the regulatory policy, but only in a partial manner. The critical cost concern should be the *total* level of costs, not whether the standard will force some of the less efficient firms in an industry to close.

The underlying difficulty with OSHA's approach is twofold. First, the framers of the legislation and the agency administrators have viewed job risks as intrinsically bad and warranting complete elimination. In contrast, I advocate that under some conditions there may be no valid reason to reduce risk levels to zero. Nevertheless, there is good reason to believe that market risks are often not at socially optimal levels; in assessing these problems, what is relevant is the nature of the market failure. The justification for policies should be whether market performance can be improved, not the level of risk per se or even whether the risk passes the test of being a "significant" hazard. To attempt, as OSHA has, to eliminate all identifiable risks is to show a reckless disregard for both the value of potentially hazardous jobs to workers and the impact of risk regulations on society.

The second major weakness in OSHA's overall design is the reliance on a rigid standards approach. Once unsafe workplaces are judged to be undesirable, the obvious solution is to require that they be made safe. Standards were used in part because of the popularity of regulatory controls, such as those used traditionally in transportation regulation, which were generally viewed as being effective policy instruments.[7] The professional bias of lawyers, who typically are relatively sympathetic to standards systems, may also have been a factor, since a disproportionate number of the leading advocates of health and safety regulations had legal training. Economists, who generally prefer tax and penalty systems to standards, did not become active in this area until after OSHA was established.

The shortcomings of the standards approach go beyond the obvious inefficiency of having an enterprise meet a requirement regardless of the particular costs and benefits of doing so. Even if one were solely concerned with workers' well-being, establishing rigid standards is unlikely to be an effective policy. Standards will enhance safety only insofar as they establish effective incentives for compliance. The policies selected by OSHA only influence these incentives; they do not otherwise alter market outcomes. By coupling stringent regulations with an ineffective mechanism for enforcement, OSHA has done little more than serve as a form of systematic harassment of the private sector.

AN OVERVIEW OF THE STANDARDS

One month after OSHA began operation, it issued over 4,000 general industry standards for health and safety, detailed in 250 pages of the Federal Register.[8] In addition, it issued extensive standards for specific industries, especially the construction trades. OSHA was able to enact these policies so quickly because it simply adopted the national consensus standards of the American National Standards Institute and the National Fire Protection Association and consolidated existing federal safety standards, including regulations such as those administered by the Maritime Safety Amendments to the Longshoremen's and Harbor Workers' Compensation Act. The reliance on industry's own standards does not imply that the standards were favorable to industry or that they were an appropriate basis for policy. The industry standards were intended as suggestive guidelines for workplace design and operation. Once they were made mandatory, the potential for major economic distortions became acute.

Indeed, the standards became the object of widespread ridicule, as OSHA regulations began to epitomize the most objectionable aspects of government intrusion in the market. OSHA was widely condemned for establishing standards for the shape of toilet seats, the precise height of fire extinguishers, and the availability of portable toilets for cowboys. One regulation requires workers on bridges to wear orange life vests so they will not drown and will be easy to locate in the water if they fall. The ineptness of OSHA's enforcement is epitomized by the fact that one company fined for violating this standard maintained that this requirement was unrelated to worker safety because the channel under the bridge had been diverted, eliminating any risk of drowning.[9]

Although the agency was required by its legislation to base its standards on existing federal regulations and national consensus standards, OSHA could have been more selective in doing so. Indeed, 928 of the most ill-conceived OSHA regulations were abolished in October 1978. The justification for revoking these rules was that they were irrelevant to worker safety.[10]

Despite the almost overwhelming volume of OSHA regulations, the agency has done very little to address the fundamental problems raised by job risks. The thousands of standards issued pertain primarily to readily visible safety hazards, the ones that market forces are best equipped to handle. The most important class of health risks, toxic and hazardous substances, is addressed almost as an afterthought in the final portion of the standards, appropriately labeled Subpart Z.[11] Almost all of OSHA's subsequent rulemaking in the 1970s was directed at rectifying this misplaced emphasis. Table 2.1 summarizes the principal OSHA initiatives in the health area between 1972 and 1981 and the present value of the costs associated with them. Two of these efforts do not pertain to regulation of specific hazards; the 1980 standard for employee access to data on their firm's hazard exposures and their medical records is a long-overdue effort to enable workers to become better informed about the risks they face, while the carcinogen policy is a fairly open-ended set of procedures for regulating cancer-related risks. If we exclude the benzene standard, which was overturned by the U.S. Supreme Court, eleven standards are listed that address specific hazards. Of these, the most costly regulations (in terms of discounted pre-

TABLE 2.1. SUMMARY OF MAJOR HEALTH INITIATIVES BY OSHA, 1972–1981

Standard	Exposure limit or description	Economic impact	Date	Status
Access to employee exposure and medical records	Employee access to records	No significant impact	7/21/78 5/23/80	proposed issued
Acrylonitrile	2 ppm limit	$298 million	1/17/78 10/3/78	proposed issued
Arsenic, inorganic	10 $\mu g/m^3$ exposure limit	$313–976 million	1/25/75 5/5/78	proposed issued
Asbestos	2 fibers/cc (present)		6/7/72	present
	1 fiber/cc (proposed)	$4.0–4.1 billion (proposed)	10/9/75	proposed
Benzene	New standard vacated by Supreme Court	—	—	
Beryllium	Engineering controls	$41 million	10/17/75	proposed
Cancer policy	Criteria for setting future regulations	[a]	10/14/77 1/22/80	proposed issued
Coke oven emissions	150 μg particulate	$3 billion	7/31/75 10/22/76	proposed issued
Cotton dust	Variable permissible exposure limits	$2.5 billion	12/28/76 6/23/78	proposed issued

TABLE 2.1. SUMMARY OF MAJOR HEALTH INITIATIVES BY OSHA, 1972–1981 (continued)

Standard	Exposure limit or description	Economic impact	Date	Status
Cotton dust (cotton ginning)	Medical surveillance	$15.6 million	12/28/76 6/23/78	proposed issued
DBCP (1,2-dibromo-3-chloropropane)	1 ppb limit	$37 million	11/1/77 3/17/78	proposed issued
Lead	Variable schedule, 200–50 $\mu g/m^3$	$660–750 million (partial)	10/3/75 11/14/78	proposed issued
Noise	85 dBA limit	$4.1 billion	10/24/74 1/16/81	proposed issued
Vinyl chloride	1 ppm limit	$561 million	5/10/74 10/4/74	proposed issued

Source: Based on calculations by the author and information in U.S. Dept. of Labor, OSHA (1980e). All figures are 1980 prices, as are all other cost and benefit estimates in this volume. Present values were calculated using an interest rate of 10 percent.

a. A range of cost estimates for the OSHA cancer policy is presented in Chapters 7 and 8.

sent value) are those for asbestos, coke oven emissions, cotton dust, and noise, each of which imposes costs on the economy in the billions.

OSHA's emphasis on perhaps the most critical type of hazards, those posing the risk of occupationally related cancer, has been disturbingly modest. The carcinogens for which OSHA initiated rulemaking proceedings in the 1970s include asbestos, vinyl chloride, coke oven emissions, arsenic, benzene, acryloni-

trile, beryllium, and a group of cancerous substances that OSHA
has dubbed the "fourteen carcinogens." Indeed, the total number
of all types of toxic and hazardous substances covered has increased very little since the initial list of OSHA standards. As a
result, OSHA standards address but a small portion of the
roughly 2,000 substances in the workplace for which there is
some evidence of carcinogenicity. OSHA's performance in the
health area has been characterized by both ill-conceived standards and a persistent disregard for the preponderence of health
hazards facing workers.

These deficiencies are interrelated. As the final column in
Table 2.1 indicates, the most recent health hazard regulation
listed, the acrylonitrile standard, was proposed in January 1978.
The moratorium on new health risk regulations during the final
three years of the Carter administration can be traced to the uncertainties raised by the benzene case and related court tests of
OSHA's authority. Challenges to OSHA's interpretation of its
legislative mandate in effect brought the regulatory rulemaking
process to a standstill. The agency's myopic commitment to an
absolute interpretation of its authority may have jeopardized the
lives of thousands of workers who could have been protected if
OSHA had regulated health risks in a more balanced manner.

Apart from these political considerations, well-designed, flexi-
ble standards can promote worker health more effectively at less
cost to society than rigid standards, which OSHA purports to be
in the worker's best interests. As the subsequent discussion will
suggest, the solution is not simply for OSHA to set looser standards. Instead, OSHA should abandon its commitment to uniform design standards and base its policies on their net worth to
society.

ALLOCATION OF OSHA'S RESOURCES

Even if OSHA standards were well designed, it is doubtful
whether OSHA's strategy for enforcing these regulations would
lead to significant improvements in workplace safety. The ineptness and inattention to the principles for effective policy which

have characterized OSHA standard setting have also been reflected in its enforcement efforts.

My intent here is not to provide a history of the agency or an in-depth view of its administrative structure.[12] An exhaustive approach of this type would be warranted only if the needed reforms of OSHA entailed only minor institutional changes. In fact, the agency's efforts are so ill-conceived that what is required is a complete revamping of its focus. In view of the extensivenes of these changes, I will focus my analysis on the pivotal elements of OSHA's enforcement strategy, which provide the most general lessons for the design of other risk regulation efforts.

As is indicated by the budgetary breakdown in Table 2.2, over half of OSHA's budget is earmarked for enforcement of the standards it sets. The remainder is allocated to such standards-related activities as standards development, compliance assistance, and program administration. With its budget of over $200 million and over 3,000 permanent positions, OSHA clearly ranks as a major regulatory agency.[13] OSHA's large budget and thousands of standards do not, however, guarantee that it enhances workers' welfare. The effectiveness of any risk regulation policy

TABLE 2.2. OSHA's budget, fiscal year 1981

Budget category	Estimate (thousands of dollars)
Safety and health standards development	8,721
Enforcement	
Federal	84,745
State	43,500
Technical support	17,547
Compliance assistance	41,689
Safety and health statistics	6,973
Executive direction and administration	8,725
Total	211,900

Source: U.S. Office of Management and Budget (1980), p. 655.

hinges on the economic incentives it creates. Since OSHA's stringent standards impose substantial compliance costs on industry, the enforcement effort must be very vigorous for firms to find it in their economic interest to comply with the program.

THE ENFORCEMENT EFFORT

The existence of mandatory health and safety standards does not in itself dictate the nature of the enforcement strategy. OSHA could, for example, penalize violations at levels reflecting the benefits of compliance, leaving it to the firm to decide whether the compliance costs exceed the health and safety benefits from changing the workplace. Instead, OSHA views the standards as rigid requirements, which it enforces through a legalistic inspection and penalty system. The three critical ingredients of the agency's strategy are inspections, criteria for standards violations, and penalties for violations. The latter two elements come into play only after an enterprise is inspected.

Ideally, one should design an inspection strategy to produce the greatest net benefits, taking into account the costs of compliance as well as the health benefits to workers. In contrast, OSHA's policy is based on the following loosely defined set of priorities: 1) inspections of imminent danger, 2) investigations of fatalities and catastrophes, 3) investigations of complaints, and 4) regional programmed inspections.[14] Despite a decade of experience with the program and considerable opportunity to learn about the value of different types of inspections, there has been virtually no change in this assignment of priorities. The only modification has been that programmed inspections now encompass what originally were two categories—special programs and random general inspections—which were formerly given fourth and fifth priority ratings.[15]

The practical consequences of the priority assignment are unclear since the actual allocation of inspections bears a strong inverse relationship to the priorities assigned.[16] Over 70 percent of inspections are the lowest-priority category of programmed inspections, which consist of general schedule inspections and follow-up inspections. The third-ranking category, complaint inspections, accounts for an additional 25 percent.

The inspection strategy is illuminated further by the federal enforcement data summarized in Table 2.3.[17] Until 1976 there was a rapid annual increase in the number of inspections, rising from under 30,000 in fiscal year 1972 to over 90,000 in 1976. In 1977 OSHA reduced the number of inspections by one-third and since then has maintained an annual level of about 60,000 inspections. This dramatic shift represented an effort to move away from the superficial inspections and penalties for inconsequential violations that had characterized its early operations.

The task of an OSHA inspector is to identify whether or not the enterprise is complying with the standards; he is explicitly prohibited from taking into account the economic costs of compliance. He must instead rely on a narrower feasibility concept tied to "the existence of general technical knowledge" that the employer could conceivably apply to meet the standards.[18]

Each subsection of a standard violated by the firm is treated as a separate violation.[19] There is, however, differentiation in the types of violations. OSHA charges a firm with a "serious" violation if the employer knows or could have known of a risk posing a substantial probability of death or serious injury or if a number of nonserious violations are grouped together and labeled serious.[20] "Willful" violations are those considered to be intentional or resulting from a failure to eliminate known risks, while "repeated" violations involve violations of previously cited sections of the standards.[21] By far the largest number of OSHA violations have been "nonserious" or "de minimis."

The changing character of violations in many respects parallels the shift in the inspection strategy (see Table 2.3). In the early years of OSHA the number of violations escalated annually, reaching 380,000 in 1976. In 1977 the number of violations dropped by more than half, and now the number totals about 130,000 annually. This dramatic shift, which occurred at the same time as the decrease in the number of inspections, was an explicit policy shift to deemphasize trivial violations, which

TABLE 2.3. CHARACTERISTICS OF OSHA ENFORCEMENT

	Fiscal year								
	1972	1973	1974	1975	1976	1977	1978	1979	1980
Inspections (thousands)	28.9	47.6	78.1	80.9	90.3	59.9	57.2	57.9	63.4
Proportion of health inspections	a	.05	.06	.07	.08	.15	.19	.19	.19
Proportion of inspections with serious citations	a	a	.04	.04	.07	.19	.26	.29	.31
Violations (thousands)	89.6	153.2	292.0	318.8	380.3	181.9	134.5	128.5	132.4
Proportion of health violations	a	.03	.02	.04	.04	.06	.09	.09	a
Proportion of serious violations	a	a	.01	.02	.02	.11	.25	.29	.34
Penalties (millions of dollars)	2.1	4.2	7.0	8.2	12.4	11.6	19.9	23.0	25.5
Penalty per violation (dollars)	23.4	27.4	24.0	25.7	32.6	63.8	148.0	179.0	192.6
Proportion of penalties for serious violations	a	a	.37	.44	.39	.61	.58	.56	.56

Source: Based on calculations by the author using OSHA computer printouts.
a. Data are not available or are not reliable.

had made the agency an object of ridicule. The impetus for these changes was not entirely OSHA's, since Congress had attached a rider to the fiscal 1977 appropriations bill requiring that no penalties be assessed for firms with fewer than ten nonserious violations.

This deemphasis of trivial violations had the expected effects on the enforcement effort. The proportion of inspections yielding serious citations quadrupled from 1976 to 1980, to reach a level of .31, and the proportion of violations labeled serious increased from .02 to .34 over that period (see Table 2.3). In part the increase in the proportion of serious violations can be attributed to the redefinition of "serious" violations, but primarily it results from a reduction in the number of nonserious violations. These figures substantially overstate the extent to which OSHA's enforcement efforts have been redirected. The nature of the change has been more of an elimination of trivial activities than a shift toward more serious hazards.

The inspections and violations are of consequence to an enterprise because they may result in penalties. A firm must be penalized for a serious violation between $300 and $1,000, depending on its gravity; nonserious violations can be free of penalty and any penalty may not exceed $300; willful and repeated violations are subject to civil penalties not to exceed $10,000; enterprises that fail to abate violations can be fined up to $1,000 daily; de minimis violations are not subject to penalty.[22]

In practice, the floors of these ranges are more relevant than the ceilings. The penalty per violation averages only $193 overall and $318 for serious violations. Nonserious violations are almost completely ignored, as they are penalized at a rate of just over $2 per violation.[23] Some OSHA officials attribute the inspectors' unwillingness to assess large penalties to the inspectors' lack of confidence in the reasonableness of the standards. Even these quite modest penalties represent a major departure. From 1972 to 1980 there was roughly a tenfold increase in the penalty per violation and in total OSHA penalties, which now exceed $25 million annually. The shift in inspection strategy was accompanied by a one-time upward shift in the proportion of penalties for serious violations, beginning in 1977 (see Table 2.3).

TARGETING OF THE ENFORCEMENT POLICY

Whether or not these enforcement efforts are socially productive depends in part on whether they are targeted to generate more efficient levels of health and safety. Perhaps the most fundamental test is to ascertain the extent to which OSHA's efforts are directed at dimly understood health risks as opposed to readily monitorable safety hazards. Every *President's Report on Occupational Safety and Health* issued by OSHA has touted health risks as the area warranting the greatest increased emphasis.[24]

The tabulations in Table 2.3 are instructive in assessing the extent to which health hazards have actually been addressed. The proportion of inspections allocated to health hazards has risen at a modest rate, reaching .08 in 1976. The relative emphasis on health apparently doubled in 1977, but .05 of the .07 increase in the proportion of health inspections was attributable to the decrease in the number of safety inspections. Since 1978 the health inspection share has remained unchanged, accounting for just under one-fifth of all inspections. In terms of its inspections, OSHA has not veered from its policy of a very gradual increase in the emphasis on health.

As one might expect, the proportion of health violations yielded by these inspections has likewise risen modestly, increasing from .04 in 1975 to .09 in 1978 and 1979. It is noteworthy that the proportion of inspections targeted at health risks is consistently more than double the proportion of health violations, indicating that the violations identified per health inspection are much lower than for safety inspections. This discrepancy is enhanced by the more time-consuming nature of health risk tests, which are considerably more complex than the visual identification of unguarded punch presses and similar safety violations. The case hours per health inspection average 43, as compared with 20 for safety, so the number of safety violations cited per hour of inspection time is more than four times as great.[25]

The safety orientation of OSHA violations is indicated by the breakdown of violations in Table 2.4. Perhaps the most striking

TABLE 2.4. DISTRIBUTION OF VIOLATIONS OF GENERAL INDUSTRY STANDARDS, 1972–1981

Category	Percentage of violations in calendar year									
	1972	1973	1974	1975	1976	1977	1978	1979	1980	1981[a]
Walking and working surfaces and powered platforms	16.1	14.7	13.8	12.9	11.7	11.2	11.5	10.6	9.5	9.5
Means of egress	4.8	5.1	5.7	6.1	5.4	4.8	4.3	4.0	3.4	4.1
Health and environmental control	2.3	2.1	1.6	1.5	1.4	1.8	2.2	2.7	2.6	2.4
Hazardous materials	6.0	7.1	7.4	8.0	8.5	8.2	8.3	8.5	7.9	8.3
Personal protective equipment	2.7	2.3	2.5	2.7	3.1	4.1	5.5	5.7	6.5	6.8
General environmental controls	3.1	2.2	2.1	3.0	3.8	2.9	2.2	2.5	2.3	1.6
Medical and first aid	1.3	1.3	1.5	1.4	1.3	1.3	1.4	1.5	1.6	1.7
Fire protection	7.5	7.3	8.3	8.6	6.8	5.2	2.7	2.4	2.3	2.5
Materials handling and compressed gas	7.7	6.6	6.4	5.5	6.3	7.1	7.4	7.5	7.0	7.7
Machinery and machine guarding	27.5	29.6	28.6	26.2	26.7	28.3	27.3	27.9	28.4	29.7
Hand-held equipment	2.3	2.4	2.4	2.2	2.2	2.1	2.1	2.1	2.1	2.2
Welding and brazing	4.7	4.5	4.3	3.9	3.8	3.8	3.8	3.8	3.7	4.1
Special industries	3.3	2.1	1.0	0.9	1.3	1.3	1.3	1.2	1.3	1.4
Electrical	9.9	12.3	13.8	16.6	16.9	16.5	17.5	16.2	15.4	13.1
Toxic and hazardous substances	0.3	0.4	0.3	0.4	0.8	1.3	2.4	3.4	5.8	4.8
Other	0.2	0.2	0.1	0.1	0.0	0.1	0.1	0.1	0.1	0.1

Source: Based on unpublished computer printouts generated by OSHA for this study. Percentages may not sum to 100.0 due to rounding error.

a. Data for 1981 are for January through May.

aspect of these patterns is the preponderance of safety violations and the consistency of OSHA's emphasis on specific hazard categories. One risk category—machinery and machine guarding—accounts for over one-fourth of all violations. This category and the following readily monitorable safety risks—walking and working surfaces and powered platforms, hazardous materials, materials handling and compressed gas, and electrical hazards—account for over two-thirds of all violations. While the relative overall emphasis on these categories has been consistent, there have been some shifts in the nature of the violations, notably the deemphasis of several safety-related categories most directly associated with less consequential violations, particularly walking and working surfaces and fire protection (such as the specific height and placement of fire extinguishers and exit signs).

Health risks have continued to receive scant attention, however. The risks that market forces are perhaps least equipped to handle—toxic and hazardous substances—accounted for fewer than 1 percent of OSHA violations through 1976 and even now are responsible for only 5 percent of all violations. (This increase is due largely to the deemphasis of trivial violations after 1977.) The absolute number of violations for toxic and hazardous substances is currently just over double its 1976 level, as compared with a sixfold percentage change. Primarily OSHA has been deemphasizing safety rather than increasing its emphasis on health and, at present, it identifies fewer than 5,000 violations annually for toxic and hazardous substances. Violations identified in health inspections tend to be in the less severe categories. Just under one-fifth of current health violations are serious, a considerable improvement from the pre-1977 period when 3 percent or fewer were in the serious category. Even with these advances, safety hazards continue to involve almost twice as many serious violations.

What these patterns suggest is that the well-known problems in monitoring health hazards may affect OSHA inspectors, who devote most of their efforts to identifying readily monitorable safety risks and, when they undertake time-consuming health inspections produce very few violations, particularly for serious risks. Since over 40 percent of the job risks cited by workers are health risks (see Chapter 4), even imperfect market treatment of these hazards may be much more effective than the efforts of OSHA inspectors.

Perhaps the only indication of a concerted effort to address health risks is the level of penalties for these violations, which rose more than tenfold from FY 1973 to FY 1979, to over $400 per violation, compared with just under $160 per safety violation. The scale of this escalation is not too dissimilar from that of the overall increase in penalties per violation (see Table 2.3), indicating that there has not been a major shift in the penalty structure so much as a continuation of the traditional practice of assessing relatively high penalties for health violations. Since total health penalties have never exceeded $5 million in any year and began to exceed $1 million only in FY 1978, one would be hard-pressed to argue that the OSHA penalty structure has provided much of a financial incentive to reduce health risks.

The allocation of the OSHA enforcement effort merits other criticisms as well. Particularly during its early operations, OSHA allocated too many inspections to low-risk industries and to small firms, where few safety improvements could be induced.[26] This allocation has improved somewhat, with inspections in small firms yielding the same number of violations per hour of inspection time as in large firms. Since more workers are affected by violations in larger firms, however, one might still question the wisdom of OSHA's strategy. Under the Reagan administration OSHA has announced a policy to exempt firms with good accident and illness records from inspections, but this solution is not ideal. Intervention should be based not on the level of risks per se but on the overall merits of intervention, particularly on the type of risk involved (for example, health risks versus safety risks).

The relative emphasis on different inspection categories also seems misplaced. Inspections based on employee complaints produce comparatively few violations, suggesting that these complaints may be motivated by concerns unrelated to safety, such as an employee's dislike for his boss. The lowest citation rate is for

follow-up inspections for previously inspected enterprises, but
this may reflect their efficacy rather than a wasteful allocation of
inspection resources. The purpose of follow-up inspections is to
provide incentives for enterprises to correct previously cited violations. If OSHA's continued interest in workplace safety is
made apparent through an active follow-up program, the return-
ing inspectors should encounter few violations.

Even if OSHA had made an effort to target its inspections
more effectively, the scale of enforcement would have been inadequate to the task. Firms' actions will be altered only if OSHA
creates effective incentives for compliance. By almost any reasonable standard, these incentives are virtually nonexistent. In
any year, the average number of inspections per enterprise covered by OSHA is under 1/50. If we exclude follow-up inspections, the chance that an enterprise will see an OSHA inspector
in any year is roughly 1/100. The number of inspections per covered worker is even smaller—under 1/1000. Since any inspection
usually addresses workplace conditions affecting many workers
rather than a single job, the proportion of workers possibly affected by OSHA inspections is actually larger, about 1/20.

Each OSHA inspection yields an average of only 2.1 violations, for which the average penalty is $193. Put somewhat dif-
ferently, the financial incentives created by OSHA now total
$7.08 per enterprise or only 34 cents per covered worker. Incon-
sequential penalty levels are reflected in even the most ambitious
cases of OSHA intervention. After the collapse of a cooling tower
in West Virginia, which resulted in fifty-one deaths, OSHA imposed $108,000 in penalties, or roughly $2,000 for each worker
killed. The evidence I will present later in this volume indicates
that the financial incentives for safety created by voluntary
market mechanisms are over a thousand times as great.

This incident is characteristic of OSHA efforts in another way
as well. Shortly before this catastrophe OSHA had inspected the
tower and cited the firm for having inadequate guards to prevent
tools from falling from the scaffold. The fundamental hazard, the
instability created by the tower's construction technique, was
completely ignored. This example epitomizes OSHA's emphasis

on trivial hazards and its failure to create effective incentives for safety.

Finally, consider the case in which OSHA imposed its largest amount of penalties on any single firm. In 1980 the Newport News Shipbuilding and Dry Dock Company was penalized $786,-000 for 617 violations. While this amount appears consequential, when averaged over the firm's 23,000 workers the penalty is only $34 per worker. Meager penalty levels such as this will do little to provide incentives to make the work environment safe.

A Framework for Assessing OSHA's Impact

The absence of an effective enforcement mechanism suggests that OSHA is likely to be ineffective in promoting worker health and safety.[27] Rather than dismissing the agency's efforts based on the structure of its policies, however, ideally one should attempt a more direct assessment of whether the agency has had any beneficial effect. To answer this fundamental question, we first must consider the way how market outcomes are determined.

The principal mechanism by which OSHA regulations affect safety is the following. OSHA policies consist primarily of specification standards for the workplace; if enforced, these standards will increase the desirability of enterprise investments in health and safety, which in turn alters the work environment. The combination of the physical environment plus workers' characteristics and safety-related actions determines the firm's health and safety performance.

The first of these links, the effect of OSHA on the physical environment, is the weakest. Enterprises will make the investments needed to comply with OSHA regulations only if it is in their financial interest to do so. Although thousands of harshly drawn standards mandate costly changes to attain the lowest feasible risk levels, OSHA's enforcement mechanism creates negligible financial incentives, so it would be quite surprising if OSHA had much effect on workplace conditions.

If the work environment improves, workers decrease their own

safety-enhancing actions, thus partially dampening any less-
ening of risk. Indeed, somewhat surprisingly, more stringent enforcement of the standards will *decrease* health and safety if the
reduction in OSHA penalties from greater investment in workplace safety exceeds the capital cost of the investment. The net
price of the investment to the firm will then be negative, leading
it to pursue such investments even to the extent that they become
counterproductive.[28] This aberrational case will not arise for the
current limited enforcement effort, so OSHA's effect on safety
should be favorable, although perhaps imperceptibly small.

The counterproductive role of worker actions is questioned by
many who believe that no worker would knowingly take an action that will increase his risk. Workers will act to minimize
risks, however, only if such efforts do not involve any tradeoffs in
terms of slower production speeds, greater effort, or changes in
health-related consumption patterns. When viewed in terms of
factors leading to more diligent risk-reducing efforts, the effect of
the work environment on behavior becomes less controversial.
Few would question that chemicals labeled hazardous are han-
dled with more caution or that workers exposed to asbestos
usually decrease or eliminate their cigarette smoking. Similarly,
if their exposure to these hazards is reduced, they will have less
incentive to enhance their safety efforts.

There have been a number of attempts to assign responsibility
for accidents on the job. Some studies have identified worker actions as a primary cause of work accidents, accounting for 45
percent of workers' compensation cases in Wisconsin, 88 percent
of accidents in a 1931 insurance company study, 84.3 percent of
British work accidents, and the majority of deaths of deep sea
divers in the North Sea.[29] Analyses making less refined distinctions have assessed the combined influence of worker actions and
workplace conditions as responsible for 63 percent of accidents
monitored by the National Safety Council and for 95 percent of
all Pennsylvania workmen's compensation cases.[30] OSHA itself,
in a detailed investigation of fatalities on oil and gas well drilling rigs, attributed half of the deaths to poor operating proce-

dures and only one-sixth to the equipment or facilities. Even by its own assessment, OSHA's exclusive focus on workplace conditions seems misdirected.[31]

Any such assignment of responsibility is intrinsically arbitrary. Since an accident or illness is the outcome of an often complex interaction of employee actions and workplace conditions, it is no more meaningful to assign responsibility for hazards than it is to allocate responsibility for other work outcomes. One cannot ascertain, for example, whether 60 percent of the worker's output is due to his equipment and 40 percent to his own efforts, since his output would be zero if either element were missing.[32]

The reason for considering worker actions is not to parcel out responsibility or to dismiss accidents as being the result solely of worker carelessness—what has been aptly termed the "nut under the hardhat" perspective.[33] Rather it is essential to understand the response of workers to regulations in order to ascertain the way in which policies exert their impact. This response is relevant whether the policy is a chemical labeling system designed to induce more careful handling, respirators that workers are supposed to wear when exposed to coke oven emissions, or punch press guards intended to diminish dismemberments. To test OSHA's impact on safety, ideally one would like to construct an econometric model reflecting these diverse influences. Such a framework could test whether observed risk levels are lower than would be predicted on the basis of other economic conditions, such as the level of employment in different industries.

Assuming that the pre-OSHA risk trend would have continued in the same manner during the post-regulation period if OSHA had not existed, simple examination of injury rate trends will provide a meaningful index of OSHA's impact. The potential pitfall of this approach is that injury rate shifts due to other economic factors may be erroneously attributed to OSHA. While injury rate trends should be interpreted with more caution than econometric evidence, they can be profitably utilized to provide a longer-term perspective on a greater variety of risk measures than do existing econometric studies.

TRENDS IN WORKER RISK

As Americans have become wealthier, risk levels of all kinds have declined. In this section I will ascertain whether OSHA has had any apparent impact on job risk trends. Examination of these trends is also instructive for assessing claims that job risks were escalating before the advent of OSHA and have since abated.

Table 2.5 summarizes the risk trends for a variety of job risk

TABLE 2.5. NONFATAL OCCUPATIONAL INJURY RATE TRENDS

Risk measure	Area	Data source	Time period	Percentage annual growth rate
Pre-OSHA				
Injury frequency rate	Manufacturing	BLS	1958–1970	+2.4
Disabling injury rate (or injury severity rate)	Manufacturing	BLS	1958–1970	0.0
Disabling injury rate	Private sector	NSC	1945–1970	−1.4
			1960–1970	−0.9
Post-OSHA				
Injury and illness rate	Private sector	BLS	1972–1979	−1.9
Injury and illness rate	Manufacturing	BLS	1972–1979	−2.3
			1974–1979	−1.8
Lost workday injury and illness rate	Private sector	BLS	1972–1979	+3.9
Lost workday injury and illness rate	Manufacturing	BLS	1972–1979	+5.0
Disabling injury rate	Private sector	NSC	1970–1980	−2.2

Source: Based on calculations by the author. Pre-OSHA calculations were based on data from National Safety Council (1979) and U.S. Bureau of Labor Statistics (1972a). Post-OSHA calculations were based on data from National Safety Council (1981), U.S. Bureau of Labor Statistics (1972b) and previous issues in that series, and unpublished BLS data. See U.S. Bureau of Labor Statistics (1981).

measures for periods before and after OSHA; Table 2.6 presents comparable data for death risks. The 2.4 percent annual increase in the Bureau of Labor Statistics (BLS) injury frequency rate for manufacturing industries from 1958 to 1970, shown in the first line of Table 2.5, played a prominent role in House and Senate deliberations and served as a major justification for the establishment of OSHA.[34] More recently this upsurge in accidents has been contrasted with the decline in risk in the 1970s as evidence of a dramatic turnabout produced by OSHA. Reliance on this measure by OSHA's proponents may not be entirely coincidental, since it is the only fatal or nonfatal risk trend that increased in the years preceding regulation.

Although there is a variety of reasons why this injury rate measure may not be meaningful, its most serious drawback is that the mix of injuries included changes substantially each

TABLE 2.6. Occupational death rate trends

Area	Data source	Time period	Death rate (percentage annual growth)
Pre-OSHA			
Private sector	NSC	1935–1970	−2.3
		1960–1970	−2.1
Manufacturing	NSC	1935–1970	−2.5
		1960–1970	−1.0
Nonmanufacturing	NSC	1935–1970	−2.2
		1960–1970	−2.2
Post-OSHA			
Private sector	BLS	1973–1979	−2.2
Private sector	NSC	1970–1980	−3.2
Manufacturing	BLS	1973–1979	0.0
Manufacturing	NSC	1970–1980	−1.2
Nonmanufacturing	NSC	1970–1980	−3.3

Source: See sources listed for Table 2.5.

year.[35] A more reliable statistical basis for assessing risk trends is an index that includes a well-defined set of injuries whose severity remains stable. In this regard, all of the other risk measures are superior. One such measure is the BLS disabling injury frequency rate for manufacturing industries (line 2 of Table 2.5), which is the severity-weighted counterpart to the overall injury frequency rate. This risk measure remained roughly stable in the pre-OSHA period. The National Safety Council's (NSC) disabling injury rate decreased substantially, particularly for long-term periods before the advent of OSHA.[36] The risk measure least subject to misclassification problems— worker death rates—also declined. All three death risk measures decreased by over 2 percent annually from 1935 to 1970. In the decade preceding OSHA, the annual rate of decrease remained roughly the same, except for manufacturing industries, which exhibited an annual decrease of 1 percent.

Although changes in the industry mix should make one cautious about placing any major emphasis on minor changes in the rate of decline in risk levels, the overall implication of the evidence is clear-cut. Contrary to widespread belief, risk levels were continuing to decline in the years preceding the establishment of OSHA. ·

Comparison of these patterns with trends after the establishment of OSHA is complicated by changes in the BLS risk measures. In 1971, after the advent of OSHA, the BLS changed the definition of injuries and switched from a voluntary reporting scheme to a mandatory system.[37] The problems created in the transitional year 1971 were so great that the BLS has never publicly released injury rate data for that year. These changes in the risk measure and in the BLS sample from the pre-OSHA period make it impossible to utilize the BLS data to gain a perfectly consistent perspective on risk trends.[38]

Despite these shortcomings, the findings for the BLS and NSC risk indices together suggest a common view of recent risk patterns. The BLS injury frequency rates for the private sector and for manufacturing declined substantially after 1972, partly because of the initially overzealous reporting of injuries arising from confusion created by the new reporting system. Since 1974 the manufacturing injury rate reflects a more modest rate of decline.

If we use the risk measure that includes only cases involving lost workdays, thus eliminating many of these classification problems, we get quite different results. Indeed, the BLS lost workday rate has increased by 4 to 5 percent annually since the advent of OSHA, both for the entire private sector and for manufacturing industries. The NSC disabling injury rate yields a quite different pattern, declining at a rate somewhat greater than in the 1945–1970 period. The ambiguities in the nonfatal risk trends suggest that the consistency and classification problems they raise limit their usefulness.

The death rate statistics, which are much more reliable, provide the weakest evidence for OSHA's efficacy. Manufacturing death rates were declining before OSHA's existence, and in the 1970s BLS manufacturing death rates remained virtually unchanged. The pattern of death rates for the entire economy continued to be almost identical to the pre-OSHA pattern.

Overall, these results undercut the purported rationale for the agency's existence and the claims regarding its efficacy. The advocates of OSHA who maintained that a "crisis in the workplace" was emerging before OSHA's existence either purposefully distorted the actual risk trends or else did not understand the deficiencies of the BLS injury rate measure.[39] These misperceptions were compounded by comparison of the alleged risk increase with more recent decreases. Neither the risk-based justification for creating OSHA nor the more recent attempts to demonstrate its effectiveness withstand close scrutiny. Workplace risks have continued to decline by about 2 percent annually, roughly the same rate as before OSHA's existence. Moreover, since the absolute level of injuries is now less, the relative invariance in the percentage rates of decline implies much smaller *absolute* decreases in worker risks since the advent of OSHA. More refined distinctions based on subtle differences among the different risk measures are not particularly meaningful due to differences in the risk samples and injury rate definitions. Similarly, one cannot ascertain whether or not other changes in the economy have obscured any possible evidence of OSHA's effectiveness in the absence of a more formal statistical analysis.

ECONOMETRIC EVIDENCE

In years in which injury rates decline, OSHA officials cite these patterns as proof of the agency's effectiveness. When risk levels rise, OSHA blames cyclical factors and shifts in the industry mix for distorting the true risk patterns. To distinguish the upbeat mythology purveyed in OSHA's press releases from quite legitimate economic factors that may affect risk levels, I will analyze the findings of econometric studies that have taken into account these diverse influences.

Unlike analysis of injury rate trends, econometric tests can distinguish the influence of OSHA enforcement policies from changes in the workforce mix, cyclical factors, and other pertinent influences. I will begin by considering the results of my analysis (Viscusi 1979b) of injury rates for sixty-one industries over the 1971–1975 period. The sample analyzed was quite comprehensive, including over four-fifths of the workers covered by OSHA.

The first issue addressed was the link between OSHA and enterprise investments in health and safety. Over the period considered, firms invested as much as $3 billion annually in health and safety, or just over 2 percent of all capital investments.[40] Since market forces would lead firms to make some health and safety investments anyway, it is not appropriate to attribute all of these investments to OSHA's influence.

Indeed, interindustry differences in OSHA's enforcement effort had no significant effect on either actual investments in health and safety in the early 1970s or such investments planned for the end of that decade. These tests analyzed the effect on such investments of past and current OSHA inspections and expected penalties, and they looked for any time trend in investments that could not be explained by economic factors unrelated to OSHA. To the extent that there has been an observable trend, it has been a decrease in the inflation-adjusted health and safety investments per worker during the 1970s.

Since classifications of an investment as being related to health and safety is somewhat arbitrary, one should be cautious in dismissing altogether the possibility that OSHA has affected firms' capital investments. Moreover, the failure to find any relationship between industry's safety investments and differences in OSHA enforcement levels does not rule out the possibility of an overall shift in investment resulting from the establishment of OSHA. The lack of investment data for the pre-OSHA period is a major shortcoming, since there may have been a shift in behavior that cannot be identified without such evidence.

This possibility should be regarded as more than conjecture. MacAvoy (1979) presents some intriguing evidence indicating that in industries most subject to OSHA inspections, prices accelerated and output growth declined compared with the pre-OSHA period. If these patterns continue to hold after adjustments are made for other economic factors, such as the extraordinarily large impact on these industries of the energy price shock, it may suggest that firms' decisions have been altered. Even if OSHA has not yet affected prices or output levels, MacAvoy's fundamental point is correct. Once OSHA standards begin to affect firms' costs, there will be a subsequent impact on their prices and output. Since the projected cost estimates for many OSHA standards are in the billions, the eventual economic impact will be considerable if compliance with the standards become widespread. Moreover, firms may alter their present investment decisions in anticipation of a possible future regulatory burden, generating an economic loss from the climate of uncertainty created by OSHA policies.

Although the absence of any evidence of a link between OSHA enforcement and enterprise decisions is a quite fundamental intermediate result, the primary matter of concern is whether there has been any influence on worker health and safety. The industry data I analyzed suggests that OSHA inspections or pen-

alties have had no significant effect on worker injuries and illnesses. This negative result pertains both to OSHA's direct effect on safety and to any indirect effect through a change in the
worker mix or in health and safety investments. There has been,
however, an unexplained drop in injuries over time, most occur-
ring between 1972 and 1973. This shift appears to be attributable
to the overzealous reporting of accidents after OSHA changed
the injury reporting system. Once industry became accustomed
to the new format, the number of reported accidents declined.
This explanation is also consistent with the health and safety investment data, which would have had to reflect a rising pattern
to show that OSHA was having a real influence on workplace
safety.

Other evidence of OSHA's impact is consistent with these results. In a study of California workmen's compensation data
from 1947–1974, Mendeloff (1979) found that after the advent of
OSHA the rates of some accidents, such as back strains, rose,
while others, such as caught-in-between injuries, declined. A
major strength of this study is that it compares injury rates before and after OSHA's existence. Nevertheless, the results were
mixed, and the instances of decreased risk levels have proved not
to be robust once more recent data are added.[41]

Although Mendeloff places greatest emphasis on those findings most favorable to OSHA, at best they indicate some possible
effect of OSHA, with no firm indication of its direction. Such
overpredictions and underpredictions of trends are quite common with statistical models of almost any economic phenomenon.
It is not appropriate to select only those instances in which injury rates are underpredicted and claim that these represent
OSHA's impact. Moreover, since none of these underpredictions
has been specifically related to any OSHA enforcement efforts, it
is unclear whether any of the observed discrepancies are attributable to OSHA, as opposed to other economic factors omitted
from the analysis.[42]

This shortcoming is not shared by R. Smith's study (1979) in
which he analyzed whether manufacturing enterprises inspected
by OSHA experienced a decrease in injury rates during the

1973–1974 period.[43] For the groups of firms analyzed, inspections resulted in a statistically significant decrease in injuries only for small plants in 1973.[44] Since the only other significant effect was an increase in the injury rates for large firms, one should be cautious in interpreting these findings, which on balance indicate that OSHA has produced a small reduction in injuries.

An even more fundamental point is that the evidence of OSHA's efficacy in this study derives solely from a decrease in injury rates at small firms, and even this minor influence may be a statistical aberration. The purpose of any test of OSHA's impact is to assess whether the underlying riskiness of the firm has decreased. The usual approach is to use the injury rate for the previous year as a measure for the firm's riskiness before the inspection.[45] This measure is quite accurate when one is dealing with very large firms or industries, but as the size of the firm decreases it becomes a less accurate index of the firm's riskiness. Consider the extreme example of a one-man firm in which the annual injury risk is .5. If the firm's employee was injured last year, the annual injury rate is 1.0, and if he was not, the injury rate is 0. High risk levels for small firms thus tend to overstate the underlying riskiness because of the dominance of such aberrational events in small samples.

This property becomes consequential when one considers that OSHA programs its inspections largely on the basis of a firm's injury and illness record. For very small firms, OSHA focuses its efforts disproportionately on those whose earlier performance overstates their actual riskiness. As a result, an apparent effect of OSHA inspections on small firms will be observed even when there is no actual impact.

In short, there is good reason to be skeptical of any evidence showing that OSHA had had an impact on job risk levels. But even if there has been some minor influence on worker health and safety, one could not conclude that the agency's efforts have been socially beneficial. That more fundamental judgment rests on whether the costs imposed in producing these gains were outweighed by the benefits of the risk reduction.

The failure of OSHA and of other risk regulation agencies suggests that direct regulatory controls are not well suited to the task of reducing risks. Many advocates of risk regulation maintain that any nonzero risk level is evidence that market mechanisms are inadequate in promoting risk reduction, and this disdain for the role of the market is quite consistent with the absolutist approach to risk regulation. In the following chapters I will advocate a quite different approach in which market forces will serve as the mechanism for promoting efficient risk levels rather than as an obstacle to be overcome. After discussing the use of market forces to control risks, I will indicate how this approach can be applied to redirect the policies of OSHA and other risk regulation agencies.

3

Compensating Differentials
for Risk

Exposure to various risks is an intrinsic aspect of many daily activities. Car travel may lead to accidents and even death, plane flights raise the risk of cancer and pose the risk of a crash, and the foods we eat create a seemingly endless variety of carcinogenic hazards. If we wished to ensure minimal risk to our lives, we would avoid all of these activities, especially those that increase our average risk exposure. Risk-seeking behavior would have the opposite effect. Our actual behavior lies somewhere between these extremes. We do not incur risks simply to endanger our lives, but we are willing to incur additional hazards in return for some offsetting advantage.[1] Participating in sports is an enjoyable form of recreation despite the risk of injury. Other risks are incurred for financial reasons, as in the case of the five hundred people who are electrocuted each year installing their own TV and CB radio antennas in an effort to avoid professional installation charges.

Workers make similar choices.[2] If a worker takes a job he knows is risky, there must be some other aspect to compensate for the risk. If the other nonmonetary aspects of the job are equivalent to those for less risky jobs, this compensation will take the form of a higher wage rate. The need to pay higher wages in turn provides a financial incentive for the employer to reduce the risk. This relationship between worker wages and the risks of the job, which is the central component of the classic economic theory of compensating wage differentials, is the principal foundation of most analyses of market-traded risk.

Although job risks have been perhaps the most heavily regulated form of risk, employment hazards are not too dissimilar from other risks that we encounter. The average blue-collar worker faces an additional annual death risk roughly comparable to that of a cigarette smoker who has four cigarettes per day or a person who rides his bike twenty-one miles per week.[3] Noise levels permitted by OSHA for work exposures are less than those encountered when we listen to a high-volume stereo, operate an electric blender, or cut the grass with a power lawnmower.[4]

Occupational hazards are not unique either in terms of their severity or in terms of the nature of the choice to face the risk. An analysis of job hazards consequently should be instructive in assessing the implications of many other risks for individual welfare. Moreover, recognition of the decisions we make daily to incur various risks should help us make more balanced judgments about policies to regulate specific kinds of risks.

ECONOMIC FOUNDATIONS

Policy debates over the appropriateness of risk regulation usually can be cast in terms of the compensating differential theory, whose fundamental economic principles were developed by Adam Smith over two centuries ago: "The whole of the advantages and disadvantages of the different employments of labor and stock must, in the same neighborhood, be either perfectly equal or continually tending to equality . . . The wages of labor vary with the ease or hardship, the honorableness or dishonorableness of employment."[5] Adherents of the market claim that this wage mechanism leads to efficient levels of job safety and to optimal matchups of jobs and workers. Critics of market outcomes at least implicitly question the underpinnings of the theory, particularly the extent to which workers actually are aware of the risks they face. These arguments can best be understood by examining the components of this theory.

Provided certain minimal assumptions are satisfied, a firm must pay a premium to attract a worker to a risky job. So long as a worker prefers more money to less and would rather be healthy than injured, ill, or dead, he will necessarily require more compensation to work at a job he believes is more hazardous. This applies to all workers, not simply those who are risk-averse, that is, those who are willing to gamble only at sufficiently favorable odds. This extra compensation may involve higher wages, greater fringe benefits, more convenient work hours, or enhanced occupational prestige. Since wages are measured in readily manipulatable units and are monitored for most classes of workers, economists can focus on the wage premium for risk, holding constant other job characteristics.

Although the theory states that workers will make a tradeoff of more money for exposure to a greater risk, one need not assume that a worker's concerns are purely financial. Moreover, the required risk premiums may be quite different for different people. Firms attempting to attract workers to hazardous jobs will only offer an adequate premium to staff those positions with capable employees. These premiums will provide a sufficient inducement for workers most willing to accept risks, while those requiring a very large wage premium will tend to select safe occupational pursuits.

If providing a safe work environment were no more costly than providing an unsafe environment, all workers would hold risk-free jobs, and no risk premiums would be paid. However, making the workplace safer typically involves substantial costs. Guards on punch presses, exhaust hoods to remove noxious fumes, and slow assembly line speeds entail either direct financial outlays or reduced output. A firm has two principal choices. It can simply pay the workers for incurrring the risks, or it can reduce its wage costs by making greater investments in the safety of the work environment. Typically these mechanisms are interrelated. The firm increases its expenditures on safety until the incremental wage reductions generated by improved safety no longer exceed the added costs of these improvements. Worker welfare enters the firm's calculations through the influence of the risk level on wages. The worker's own valuations of risk in effect determine the price the firm must pay if it does not find it financially worthwhile to diminish the risk.

The interaction of these decisions by workers and firms determines the number of people who are exposed to different levels of risk and the premiums these hazards command. The outcomes are not completely capricious and in many instances are not too dissimilar from other market processes. To characterize the risks facing firemen, those in the military, and professional athletes as being essential to these pursuits, while at the same time viewing as unnecessary the black lung risks faced by coal miners, is to neglect the common economic elements in these situations. If the market is functioning properly, the level of risk will reflect both the costs of ameliorating the hazard and the benefits to workers of doing so. Differences in technologies and in individual attitudes toward risk will result in a variety of risk levels in the economy, but this disparity in no way implies that some risks are less necessary than others.

Ever since the carnage in the workplace during the Industrial Revolution, critics have challenged the validity of the compensating differential theory, citing the large numbers of occupational injuries and deaths as *prima facie* evidence of inadequacies in the market. Except in extreme instances, however, the absolute level of the risk tells us very little. Is the 1/10,000 annual risk of death that faces U.S. workers too high, too low, or just right? Certainly some nonzero risk level is optimal, and we would undoubtedly consider it catastrophic if the risk were 1/100 or 1/10. However, fine gradations at the low levels of risk incurred by most workers provide little basis for judgment.

The most meaningful approach to assessing market deficiencies is to examine the ways in which the risk premium analysis fails to reflect the inadequacies of functioning labor markets. A particularly lucid critique was offered in Engels's *The Condition of the Working Class in England*. Engels not only examined the health and safety hazards in detail but also identified pivotal shortcomings of the market—the inadequacy of compensation for injured workers, the lack of worker education and knowledge concerning the risks, and falsification of accident records by mill owners to present an overoptimistic portrayal of conditions at their firm.[6]

These themes overlap with the three critical ingredients of the compensating differential analysis. First, the theory assumes that workers are fully cognizant of the risks and their implications. In many instances, the worker does not have complete information about the risks, and even if he does, he may not fully understand the implications of the risks for his welfare. Particularly for carcinogenic hazards with time lags of a decade or more before a health problem is observed, there may be considerable difficulty in assessing the implications.

Market outcomes also will be optimal only if the insurance arrangements function in an effective (actuarially fair) manner.[7] The shortcomings of private insurance have contributed to the establishment of a comprehensive workers' compensation system. Some of these inadequacies, such as adverse selection (meaning that only the bad risks purchase insurance), are problems associated with the behavior of workers and their employers rather than deficiencies of the insurance industry.

The final critical feature of the compensating differential theory is the assumption that the worker's valuation of the health risks fully reflects the value of the risk to society as a whole. This assumption would be violated if, for example, other people had a major altruistic concern with the worker's well-being. To the extent that the worker takes the preferences of other family members into account when determining the premium he requires to accept the risk, these concerns will be reflected at least in part in labor market outcomes.

The special status accorded to health-enhancing government programs suggests that society at large also has an interest in workers' well-being. Unless this altruistic concern is especially strong, however, its effect is outweighed by the value the worker himself places on his health. Moreover, job risk regulations imposed by society may accord with the preferences of policymakers but not with those of the workers whose jobs are being regulated.

Job hazards also have important and more tangible effects on other members of society. The job hazard may be part of a larger environmental risk problem. Operations at the Rocky Flats nuclear power plant in Denver, for example, may have increased the risk of cancer in people living downwind from the plant. In addition to these health effects, job risks may have substantial financial implications for society, particularly with established systems of social insurance and workers' compensation, whose costs are shared by all.

Although there are clearly major deviations from the smoothly functioning world of the compensating differential theory, one should not dismiss the legitimate insights of the analysis. Many of the theory's shortcomings are difficult to quantify, so one cannot ascertain whether the inadequacies of the analysis are critical or of only technical interest. Moreover, even in the case of clearly important deviations from the classical analysis, such as the lack of perfect information, complementary market responses may work to promote the desired outcomes.

An Assessment of Risk Premiums

Advocates of government regulation of risks typically do not believe there is widespread, significant compensation for job risks. Risk premiums are presumably limited to special cases, such a professional athletes, for whom the presence of risk compensation is well known. As former New York Jet quarterback Joe Namath observed in discussing his permanent knee injuries, "We got paid pretty good. Nobody was twisting our arms."[8] But professional athletes are not only the workers who receive risk premiums. In some cases, the level of hazard pay is specified precisely in contract provisions. Elephant handlers at the Philadelphia Zoo receive an annual premium of $1,000 because "elephants will work only with people they like, and if they don't like them, the handlers face extra risk."[9] Formal specification of risk premiums is not the norm, however, as the discussion below of collective bargaining agreements will indicate. More typically, job hazards are a component of a job evaluation system that gives each job a rating, which in turn affects the wage that is paid.

The most meaningful test for the presence of risk premiums is to utilize conventional statistical techniques to assess the incremental effect of risks on worker earnings, taking into account other determinants of wages such as the worker's education and experience, the type of job, and regional economic conditions. A large number of studies of risk premiums have followed this approach, and the consensus view is that workers who incur risks are paid substantial rewards. In the discussion here I will focus on the results of my study of wage premiums for a group of 496 blue-collar workers.[10] Chapter 6 provides a more comprehensive review of related studies, which have led to sufficiently similar results that we should be quite confident of the general economic importance of risk premiums.[11]

Studies of risk premiums should be interpreted with some caution, however, since none of them has been ideal. The theory addresses the presence of premiums for risks perceived by workers. To measure these premiums, one would like to have information regarding workers' perceptions of the probabilities of particular types of risk, such as death, injury, or illnesses of differing severity. Instead, the only available data pertain to workers' perceptions about whether their jobs expose them to dangerous or unhealthy conditions. Although this risk measure captures the importance of subjective beliefs, it does not distinguish different gradations of risk.

An alternative approach is to use average risk measures, such as average injury rates, for the worker's industry or occupation. Although this index captures differing degrees of risk, it does not necessarily reflect the worker's beliefs about the risk of his particular job, which is what is relevant for the analysis. Moreover, if workers understand the industry-wide risk but not the risk posed by their own jobs, we will observe wage premiums for the industry even though the worker may substantially misassess the hazards of his job.

Since subjective and objective measures have differing relative strengths and weaknesses, we should explore the implications of each risk measure in order to assess the robustness of the risk premium estimates. In my study of workers' subjective risk perceptions, I found that workers who believed that they were exposed to dangerous or unhealthy conditions received over $900 annually (1980 prices) in hazard pay.[12] It is especially noteworthy that an almost identical figure was obtained when I used an objective industry injury risk measure as the risk variable. The similarity of the findings using subjective and objective measures of risk lends strong empirical support to the validity of the risk premium analysis.

Unfortunately, these results do not enable us to conclude that markets work perfectly. Is the premium less or more than would prevail if workers and employers were fully cognizant of the risks? The size of the premium only implies that compensating differentials are one element of market behavior. A more meaningful index is the wage premium per unit of risk. If it is very likely that a worker will be killed or injured, a $900 risk premium can be seen as a signal that the compensating differential process is deficient. The average blue-collar worker, however, faces an annual occupational death risk of only about 1/10,000 and a less than 1/25 risk of an injury severe enough to cause him to miss a day or more of work. Consequently, the observed premium per unit of risk is quite substantial, with the implicit value of life being on the order of $2 million or more for many workers (see Chapter 6).

The safety incentives created by market mechanisms are much stronger than those created by OSHA standards; a conservative estimate of the total job risk premiums for the entire private sector is $69 billion, or almost 3,000 times the total annual penalties now levied by OSHA.[13] Whereas OSHA penalties are only 34 cents per worker, market risk premiums per worker are $925 annually. This figure would be even higher if we added in the premiums that are displaced by the workers' compensation system, which provides an additional $11.8 billion in compensation to workers.[14]

Another pertinent question is whether there are different financial rewards for different kinds of risks. It is often suggested that health hazards do not command risk premiums, whereas safety hazards do. However, for the subjective measure of health risks perceived by workers, there are no statistically distinguishable differences between compensation for health risks and that for safety risks.[15] These findings do not relate to rewards for health hazards that are not fully known. Nevertheless, it is clear that the compensating differential process is not restricted to safety risks alone.

INDIVIDUAL WEALTH AND RISK PREFERENCES

Despite the intuitive appeal of the compensating differential theory, its validity has long been challenged. One of the chief criticisms was voiced by John Stuart Mill, who regarded it as somewhat paradoxical that the most attractive jobs in society are also the highest paid. This pattern appears to contradict the compensating differential theory, since the riskiest jobs presumably should have the greatest rewards.

Such seemingly aberrant behavior can be reconciled with the compensating differential theory by considering the influence of one's wealth status on attitudes toward risk. Individuals' willingness to incur risks will vary with a variety of personal characteristics, including age, sex, number of children, education, and individual tastes. The characteristic of particular concern here is financial status. Assuming reasonable patterns of individual preferences, a worker with greater wealth will be less willing to incur job risks or, viewed somewhat differently, the premium necessary to induce him to accept any particular risk will be greater.[16]

This behavior is similar to many other patterns of consumer choice. Richer consumers purchase better cuts of meat, more comprehensive health insurance, and higher-quality cars. The influence of a worker's wealth on his willingness to incur an occupational risk arises from a similar variation in tastes. Differences across income classes in employment opportunities and job risk information will tend to reinforce this pattern.

Individuals at the top of the occupational hierarchy typically have higher levels of lifetime wealth than those at the bottom. Moreover, since they are better educated and have more marketable skills, they have a wider range of work opportunities. Their more affluent economic status will be reflected in a lower willingness to boost their income even further through work on a hazardous job, so premiums for risk account for little of the income of the richest workers.

Lower-paid workers differ from those at the top in two principal ways. First, even if these individuals took the riskiest and most unpleasant jobs available, they would not be able to earn as much as those at the top of the income scale. Second, their greater willingness to accept risks in return for financial compensation has boosted their income status above what it would otherwise have been. If all jobs were required to be as safe as the most highly paid white-collar positions, the income status of those at the bottom of the income scale would be lowered further. Wage premiums for risk do exist, but they are not sufficient to offset all of the other factors generating the low-income status of the workers who receive them.

Consideration of wealth effects enables one to resolve the apparent paradox that the most attractive jobs in society are also the highest paid. Any person will require a premium for taking a job with a greater perceived risk. However, the confounding of variations in work opportunities for people having different income status (because of educational differences, perhaps) with risk preference variations arising from differences in wealth makes it impossible to distinguish the role of compensating differentials based on a broad overview of income patterns.

To assess the effect of individual wealth on the risks incurred, ideally we would like to avoid the complicated interrelationship between a worker's wealth and other attributes that influence his earnings. Controlling for these factors, I have found that wealthier workers were employed in safer jobs. More specifically, there is a negative relationship between a worker's net assets and the average risk in the industry in which he works.[17]

The importance of the wealth-risk link extends beyond its usefulness in technical analyses of risk. The pivotal determinant of many observed risk patterns often can be traced to variations in financial resources. Throughout this century there has been a steady decline in the accident rate associated with various activities. As we have become increasingly affluent, we have become less willing to incur risks. This pattern is shown quite dramatically by the fatal accident data in Table 3.1 (nonfatal accident data reflect similar patterns but are more subject to problems of classification). The bottom line in the table summarizes the steadily increasing value of per capita income. Adjusted for inflation, real incomes have almost tripled in the past forty years, while accident rates have declined by a more modest amount. The percentage annual growth rate in per capita income, 2.6 percent, is not too dissimilar from the annual rates of decline in many categories of fatal accidents. Total accidental death rates have been declining by just over 1 percent a year; excluding motor vehicle deaths, fatal accident rates have decreased by almost 2 percent annually; and job-related death rates have been decreasing by over 2 percent annually. Fatal accident rates in all but two categories (motor vehicle deaths and total accidental deaths) declined continuously throughout this period. The two aberrational patterns can be traced to the increase in auto accident rates from 1960 to 1970.

The inclusion of motor vehicle death rates distorts the relative safety of different activities since there has been an increase in the number of drivers and the average miles. A more meaningful index of automobile safety, which isolates the riskiness of the activity, is motor vehicle deaths on a mileage basis, and these have declined steadily. Changes in the amount of automobile travel should be taken into account, since travel mileage has increased for reasons quite apart from any shift in risk preferences. Automobile technology has changed, as have residential patterns and highways. There have also been rather stark changes in the age structure and size of the driving population. The dramatic increase in the number of high-risk teenage drivers in the 1960s was due to demographic factors, notably the postwar baby boom, not to any change in the risk of automobile travel.

The other categories in the table are less affected by changes in the level of the activity and consequently better reflect the increase in safety. Accidents both at work and at home have de-

TABLE 3.1. FATAL ACCIDENT TRENDS

Accidental death category	Accidental death rate						Relative percentage change, 1940–1980	Percentage annual growth rate, 1940–1980
	1930	1940	1950	1960	1970	1980		
Total (per 100,000 population)	80.5	73.4	60.3	52.1	56.2	46.2	−37	−1.2
Total excluding motor vehicle	53.8	47.3	37.3	30.9	29.4	23.0	−51	−1.8
Work (per 100,000 population)	15.4	12.9	10.2	7.7	6.8	5.7	−56	−2.0
Work (per 100,000 workers)	—	38	27	21	17	13	−66	−2.6
Manufacturing	—	19	17	10	9	8	−58	−2.1
Nonmanufacturing	—	43	31	25	20	15	−65	−2.6
Motor vehicle (per 100,000 population)	26.7	26.1	23.0	21.2	26.8	23.2	−11	−0.3
Motor vehicle (per 10^8 miles)	15.6	11.4	7.6	5.3	4.9	3.5	−69	−2.9
Home (per 100,000 population)	24.4	23.9	19.2	15.6	13.2	10.1	−58	−2.1
Public nonmotor vehicle (per 100,000 population)	16.3	12.5	9.9	9.4	11.5	9.2	−26	−0.8
Per capita income (1972 dollars)	1,767	1,912	2,624	3,085	4,228	5,348	+180	+2.6

Source: Based on calculations by the author using data from National Safety Council (1981) and U.S. Department of Commerce (1980) and earlier issues. The implicit GNP deflator was the price measure used.

clined by 2 percent annually. The relative risk at work compared to the risk at home is of special interest because of the widespread claim that employees are safer at work than at home. While over the entire population there are clearly more home accidents than work accidents, this is not an appropriate basis of comparison, since not all people work. Fatal accidents per worker exceed the fatality rate per person at home.

Perhaps more important is any change that may have occurred in the relative safety of home and of work. Have activities at work become comparatively safer than those at home because of improved technology, the rise of the services sector, and increased awarenes of job risks? The decrease in the relative riskiness of work is quite modest and may be accounted for solely by the decline in work hours. For the 1950–1980 period for which overall worker hours information is available, the ratio of fatal home accidents per person to fatalities per worker rose by 10 percent. Since average work hours declined by almost an identical amount over that period, the decrease in time at work explains any apparent change in relative riskiness.[18]

As society has become richer, there has been a widespread decline in accidents both at home and at work. Although the general trend is clear, the particular mechanism of influence is not. Two types of effects are most instrumental. First, increased wealth and the associated decreased willingness to incur risks may change people's choices. A richer worker may choose a safer job, firms in a more affluent society will adopt less risky technologies, and wealthier consumers will be more likely to buy safer cars. A second effect is that, especially over long periods of time, increased wealth alters the options available. Wealthier societies have a greater incentive to develop airbags, more durable products, and other technological innovations that decrease the risk of particular activities. The extent to which the wealth-risk relationship arise because greater wealth induces a change in technology or a change in choices within a technology cannot be ascertained. For our purposes, disentangling the components of the mechanism of influence is less important than understanding the overall relationship.

Wealth differences will lead to variations in the hazards incurred among different population segments as well as over time. Table 3.2 divides the states into five quintiles, ranked by per capita income. The relationship of per capita income to the rate of fatal accidents is remarkably consistent. The two principal categories are total fatal accidents and fatal auto accidents on a mileage basis. In each, the risk level increases as the income rank of the states declines. The components of the accident rate display roughly similar effects, although the patterns are not as regular.

One would expect to find similar differences across countries. More affluent nations, such as the United States, should choose safer technologies and display greater risk-avoiding behavior than less developed countries. Available international accident data, however, yields the opposite relationship, with risks and per capita income positively correlated.[19] One reason for this de-

TABLE 3.2. STATE ACCIDENT LEVELS BY STATE INCOME LEVEL, 1978

Group	Per capita income	Fatal accidents per 100,000					Mileage auto accident rate[a]
		Total	Work	Home	Public nonmotor vehicle	Unclassified	
1[b]	8,720	38.9	c	c	c	c	3.0
2	8,079	42.4	3.0	7.3	10.0	0.8	3.2
3	7,568	46.6	3.1	8.9	9.5	2.0	3.2
4	6,891	59.8	4.7	10.4	11.3	2.4	3.8
5	6,320	60.5	3.5	9.9	12.1	2.4	4.0

Source: Accident data for the states are from National Safety Council (1979). Per capita income figures are based on data from U.S. Department of Commerce (1980).

a. Mileage death rates pertain to deaths per 100 million vehicle miles.

b. Nevada, Wyoming, and Alaska, which are among the richest and most hazardous states, are excluded from Group 1 because their aberrational income and risk patterns are attributable to their unique economic and geographic conditions. Nevada and Alaska are at the extreme in terms of both risk and income.

c. Figures are not given for these components because data are unavailable for the majority of the population covered by these categories.

parture from expected behavior is the differences in individual activities in more and less developed countries; the high auto accident death rates in the United States, Austria, and West Germany are principal contributors to these countries' relatively high accident rates. A second factor is that in advanced countries the mortality rate from illnesses (adjusted for demographic mix) is lower, so the proportion of healthy individuals who might be accident victims is greater. Finally, and perhaps most important, the reporting of accidents is more meticulous in more affluent countries because of greater sensitivity to risk and individual health. Four countries ranked among the safest on the basis of reported fatal accidents are the Dominican Republic, Singapore, Hong Kong, and Chile. In contrast, the advanced countries of Western Europe and the United States have very high accident rates even when motor vehicle deaths are excluded. Differences in record keeping are perhaps the most plausible explanation for these somewhat surprising patterns.

The more affluent countries are characterized by a greater emphasis on policies to reduce risks, as one would expect. As the United States has become richer, the social regulation of risk has escalated. Earlier in this century risk regulations were confined primarily to standards for food processing and drug safety.[20] The same shift in attitudes toward risk that led to the proliferation of risk regulation agencies in the 1970s has also contributed to the influence of Ralph Nader and to a variety of risk-oriented efforts, such as the movement to promote the safety of nuclear power plants. These developments have been interdependent; for example, Ralph Nader's influence has contributed to the public sensitivity to hazards. In the absence of the effect of greater affluence on our willingness to incur risks, however, it is unlikely that these political efforts would have met with much success.

A potential pitfall for policy design is that the individuals who wield the greatest influence over the structuring of policies tend to be richer than those who are exposed to the risks. The disparity in wealth between those who formulate policies and those affected by the policies could provide the impetus for regulations that would reduce the risk exposures of broad segments of the

population. If the need for regulation is based simply on a difference in risk preferences arising from disparities in wealth, it is
inappropriate for the government to regulate the risks. To be
sure, policy choices usually compound concerns arising from differences in wealth with more compelling rationales for intervention, such as inadequate risk information. However, if differences in risk preferences are the primary determinant of public
policies, the resulting regulations will be especially misguided. If
coke-oven workers are willing to endanger their lives in return
for substantial salaries, or if India chooses to develop nuclear energy as the most promising energy source for its long-term development, government efforts to interfere with these decisions will
reduce the welfare of those whose choices are regulated. An ethical issue that invariably arises in these instances is the extent to
which policymakers are simply imposing their risk preferences
on others.

Among the more extreme manifestations of this interventionist mentality are proposals to impose risk standards on activities
in foreign countries that have no direct effect on the risks incurred by U.S. citizens.[21] Some analysts have urged a ban on im-
portation of unsafely produced goods. A broader variant is the
suggestion that uniform standards for health and safety be developed and applied internationally. Policies of this type are
fundamentally ill conceived. The profound differences in wealth
and economic development across countries necessarily lead to
quite disparate attitudes toward risk. Penalizing workers in less-
developed countries by not buying their products will not boost
their welfare.

More generally, insistence on uniform hazard regulations will
inevitably lead to the types of compromises that are detrimental
to all concerned. More affluent countries, such as the United
States and the industrial European nations, would prefer
tighter standards than those reached by international consensus,
while less-developed nations would prefer looser standards. The
inherent shortcoming of standardization is that uniformity is
not desirable. Variations in financial resources give rise to quite
valid differences in risk preferences that should not be overrid-
den. The greater danger from wealth differences is not that the

poor will choose to incur risks, but that the rich will take interventionist actions.

Wealth-related concerns are also at the forefront of policies to regulate the riskiness of U.S. exports. Should the State Department ban the export of aerosal sprays, Tris-coated sleepwear, or pesticides that are banned in this country? Many of the countries that would receive these goods are poorer and at a stage of economic advancement not too unlike that of the United States before the advent of most risk regulations. Banning products for export may diminish the welfare of these people, impede their economic development, and even result in decreased longevity as resources are reallocated in response to the ban.

A more appropriate basis for export restrictions would be linked to clearly legitimate economic concerns. If the United States does not impose safety standards on its exports, overseas consumers of American products may expect that the same standards are applied to exports as to domestic outputs, thus leading them to misperceive the risk. Moreover, without export restrictions, U.S. producers and the U.S. government may be culpable. If an American-made nuclear reactor that is not subjected to U.S. safety standards leads to a major catastrophe, people in foreign countries may reject U.S. products in general. This response is not simply a political matter. A major change in foreign consumers' assessments of the risks of U.S. goods could have profound implications through market forces alone.

Legitimate economic influences such as these should be mustered to provide a rationale for export restrictions and other regulatory actions. What is important is that these motivations be made clear so that differences in attitudes toward risk do not dominate the policy process.

THE ROLE OF UNIONS

The wealth-risk relationship is also a fundamental determinant of labor union actions, since the wealth of union members has a pivotal effect on unions' behavior.[22] Unions have traditionally played a major role in promoting occupational safety and in providing political support for risk regulations. Most recently, union officials have been outspoken in their claims that we should not trade off dollars for lives and, as a consequence, that governmental risk regulations should be set at the most stringent feasible levels. I will present evidence suggesting that unionized workers do indeed trade off dollars for added health and safety risks, though at a much higher price than other workers.

It has been well established that unions raise worker wages by roughly 10–15 percent. Of interest here is the composition of this increase. More specifically, do unions alter the tradeoff between financial rewards and risk, and what factors have contributed to this emphasis? There are two principal reasons why wage premiums might be set differently for unionized than for nonunionized workers. First, unions may possess better information about job risks than do individual workers. The relative permanence of unions and their coverage of broad groups of workers enables them to acquire more diverse kinds of information about job risks. The Oil Chemical and Atomic Workers union, for example, has developed considerable expertise in occupational health and safety issues.

A second, more fundamental reason stems from the nature of unions' representation of workers. In normal competitive markets, it is the marginal workers—the potential new hires and those on the verge of quitting—who are most instrumental in influencing the level and composition of the pay package because it is the actions of these workers that will be altered by the firm's decisions. The other employees, usually referred to as the "inframarginal" workers, are less sensitive to the firm's decisions, often because of their substantial experience and seniority in the firm. Since unionized workers are tied to the firm by seniority rights and pensions much more than their nonunion counterparts, an important policy issue is whether these relatively immobile workers are trapped in hazardous jobs they would like to leave or whether unions protect their interests in some way.

When a union bargains on behalf of workers, its focus is usually on the preferences of all workers, not simply those of the marginal employees. This emphasis has been borne out in a series of recent studies relating to different aspects of compensation, including fringe benefits and hazard premiums.[23] The inframarginal workers generally are older, more experienced, and wealthier. This greater affluence decreases their willingness to incur risks, so these workers will demand a greater premium per unit of risk. If unions advance the interests of these relatively immobile employees, the market outcome will be more efficient. A possibly beneficial role for unions is not consistent with the standard compensating differential analysis, since that framework does not view workers as being tied to the firm through their firm-specific skills or other factors.

In many cases, unions' concern with job risks is reflected in formal contract provisions. The breakdowns in Table 3.3 indicate that most union contracts provide for compensation of job-related injuries, most frequently supplemental pay for time not worked. A smaller proportion of contracts, covering one-eighth of all unionized workers, contain explicit provisions for hazardous duty differentials. Almost all union contracts make explicit provisions for the work environment as well (see Table 3.4). Although the most common provision is a general commitment to workplace safety and health, contracts usually specify the responsibilities of the employer and the worker and also provide for safety equipment, first aid and hospital facilities, physical examinations, and safety inspections.

The formal treatment of job risks in contracts does not necessarily mean that unions alter relationships that would otherwise have prevailed. Since one would expect job risks to affect worker wages even for unorganized workers, the contract provisions may simply formalize the kinds of outcomes that would be observed in the absence of the union.

The incremental effect of unions can be distinguished through a modification of the standard statistical test for wage premiums. Available evidence suggests that the preponderance of wage premiums for risk are for unionized workers. Unions alter the wage-risk tradeoff dramatically, boosting the monetary tradeoff by over 50 percent above the average for workers overall.[24] Although the limitations of available data should lead one to be

Table 3.3. Collective bargaining provisions affecting job risk premiums, 1974–75

Provision	Agreements with such provisions	Workers covered (thousands)
Hazardous duty differentials	260	1,005
Falling risks	161	575
Excessive heat or fire	21	77
Radiation hazards	12	31
Electrical work	13	25
Acid, fumes, or chemicals	109	373
Explosives	42	250
Compressed air	80	294
Unable to determine	4	31
Compensation for job-related injuries	1,038	4,655
Temporary continuation of wages	634	2,905
Supplemental pay for time not worked	710	3,339
Special transfer provisions (red circle rate)	83	251
Total sample size	1,724	7,878

Source: U.S. Bureau of Labor Statistics, *Major Collective Bargaining Agreements: Safety and Health Provisions,* (1976), pp. 63–64.

cautious in drawing precise conclusions regarding the risk premiums received by different groups of workers, it is well established that unionized workers do receive larger risk premiums.

If forced to pay greater premiums for risk, a firm will find it attractive to increase its safety-enhancing expenditures to bring about a lower level of risk. The evidence on the level of risks by union status is, however, ambiguous. Although unionized workers face lesser health risks, they face greater safety risks than unorganized workers and incur roughly the same overall risks.[25]

TABLE 3.4. COLLECTIVE BARGAINING PROVISIONS AFFECTING JOB HAZARDS, 1974–1975

Provision	Agreements with such provisions	Workers covered (thousands)
General safety and health provisions	1,607	7,197
Union-management cooperation on safety	757	3,946
Safety committees	567	3,222
Employer compliance with health laws and regulations	585	2,666
Employee compliance with safety and health rules	805	3,283
Employee discipline for noncompliance	349	1,522
Safety inspections	335	2,357
Safety equipment	847	4,295
Safe tools, equipment, and materials	164	783
Crew-size safety regulations	222	870
Sanitation, housekeeping, and personal hygiene	673	2,994
Physical examinations	554	2,963
Accident procedures	459	2,746
First aid and hospital facilities	492	3,020
Specific hazard protection		
Adequate lighting	104	564
Eye and face protection	338	1,427
Protection from noxious gases and dust	285	2,218
Protective clothing	367	2,373
Gloves	297	1,569
Safety shoes	168	462
Hearing protection	67	1,286
Falls	146	777
Falling objects	174	705
Machinery guards and safety devices	115	440
Protection from electrical hazards	80	305
Radiation	72	520
Fire protection	188	1,007
Hazard to fellow employee	128	356
Total sample size	1,724	7,868

Source: U.S. Bureau of Labor Statistics (1976).

Since unions have focused on the more hazardous firms in their organizing efforts, these patterns may be consistent with a reduction in risk levels at unionized firms. The net effect of unions on worker risk cannot be fully resolved without a detailed analysis of risk and compensation levels before and after unionization.

In situations in which there are divergences from the assumptions of the compensating differential theory, a well-designed regulatory policy will call for complex and often infeasible calculations of the merits of different alternatives. Perhaps the most powerful implication of the compensating differential analysis is that if its principal assumptions are satisfied, market forces will set job risk levels efficiently on a completely decentralized basis, involving no interference with the decisions by workers or firms. Even in situations in which there are obvious deficiencies in market processes, the risk premium mechanism usually serves a productive economic function in providing incentives for safety and in allocating workers to jobs in a reasonably efficient manner.

4

MARKET FORCES AND INADEQUATE RISK INFORMATION

THE PRINCIPAL LIMITATION OF THE COMPENSATING DIFFEREN-
tial process is that appropriate risk premiums and efficient
matchups of jobs and workers may not result if workers are not
fully aware of the risks they face. Some job risks are widely
known and should generate risk premiums. In the usual situation, however, the exact probabilities of being killed or injured
are not known by anyone. Owing to the retarded state of occupational medicine, even the underlying medical ramifications of
different exposures to hazards of the workplace such as radiation, noise, high temperatures, and chemical vapors are not well
understood. This uncertainty is often compounded by lack of
knowledge about characteristics of the work situation—for example, the concentration of asbestos fibers in the air. Adverse
health effects that occur very infrequently and are not apparent
for decades create especially severe difficulties, since it is often
not possible to ascertain the cause of the ailment.

Market critics often cite these shortcomings as evidence that
market processes do not work and that risk regulations are
needed. The view I will take here is less extreme. Workers are
aware of many job risks, and for these the risk premium mechanism is relatively effective in promoting efficient levels of safety.
Workers who are not fully cognizant of the risks associated with
a job often can learn more about these risks once they begin work
on it. If this information is sufficiently unfavorable, given the
wage they are paid, they will quit The job hazard–quit relation-

ship, which has been the focus of much of my research, can be viewed as an extension of the market responses to situations in which workers can learn about risks through their work experience. The quitting response reduces the problems associated with workers having inadequate information about the job risks they face, but it may not eliminate the fundamental difficulties.

ARE RISK PERCEPTIONS ACCURATE?

The fundamental issue involved in any assessment of the classic model of compensating differentials for risk is whether or not workers are aware of the risks they face. Most workers have many sources of information for making some judgment about these risks. Before working on the job, they can use information about the firm's reputation, the nature of the job, or the experiences of friends who have worked there. Some risks, particularly newly discovered carcinogenic hazards or the black lung problems of coal miners, have been highly publicized and are familiar to the general public as well as to the workers themselves. Once he is on the job, the worker can observe the workplace conditions and the effects of the job on his well-being and that of his coworkers.

These diverse sources of information about job risks are reflected in workers' risk perceptions.[1] Over half of all blue-collar workers surveyed believe that their jobs expose them to dangerous or unhealthy conditions. Although this result contradicts the widespread belief that workers are ignorant of the risks they face, it in no way implies that they accurately perceive the risks cited or that they are aware of all of the hazards posed by their jobs. Table 4.1 gives a breakdown of the hazards cited by workers, listing the first hazard cited by the worker and the total frequency with which the hazard was cited. The hazards mentioned most often were those posed by inherently dangerous materials (chemicals, gases, smoke, and fumes), inherently hazardous equipment, inherently hazardous procedures, placement hazards, and dangerous exposures to dust and other materials. Detailed examination of these hazards reveals that the risks are

TABLE 4.1. DISTRIBUTION OF JOB HAZARDS CITED BY WORKERS

Nature of hazard	Frequency as first hazard cited	Overall frequency of citation of risk
Inherently dangerous materials	.1008	.1734
Inherently hazardous equipment	.0786	.1068
Inherently hazardous procedures	.0625	.0927
Dangerous exposure to dust	.0383	.0584
Transportation hazards	.0282	.0363
Placement hazards	.0242	.0726
Exposure to communicable diseases	.0202	.0262
Temperature or humidity extremes	.0202	.0464
Exposure to elements	.0181	.0301
Inadequate help	.0141	.0242
Violence from customers	.0141	.0242
Inadequate safety procedures	.0121	.0201
Other exposures to violence	.0121	.0161
Defective machines or equipment	.0081	.0222
Inadequate shoring	.0081	.0141
Slippery floors or footing	.0060	.0221
Poor sanitation	.0060	.0120
Excessive noise	.0060	.0060
Inadequate guards on machinery	.0040	.0080
Inadequate protective equipment or clothing	.0040	.0060
Inappropriate job for physical capabilities	.0020	.0040
Inadequate hazard warnings	.0020	.0020
Danger from exposure to animals	—	.0020
Inadequately guarded electrical apparatus	—	.0020
Other illness or injury	.0081	.0141

Source: Based on Viscusi (1979a), Table 14.2.

broadly consistent with the particular job. Temperature and humidity extremes were cited by a truck driver for a canning company, inadequate shoring was listed by a construction worker, and slippery floors and footing was cited by a manufacturing worker in the plastic products industry.

A distinction is often made between health risks and safety risks. Although the distinction is somewhat arbitrary, health hazards typically refer to environmental exposures that lead to internal health problems. I have classified the following categories in Table 4.1 as health risks: radiation, inadequately labeled chemicals, dangerous materials (such as fumes), dangerous exposures to dust, poor sanitation, excessive noise, temperature or humidity extremes, exposure to the elements, and exposure to communicable diseases. The other hazards, such as inadequate machine guards, were labeled safety risks. These typically pose more immediate and visible external threats to the worker's well-being and are usually associated with accidents rather than illness.

It is standard to characterize workers as being aware of safety hazards but ignorant of health risks. But such sweeping statements are not borne out by the responses of workers, since two-fifths of the hazards cited are health risks. Although many health risks, such as those posed by the bewildering array of toxic substances, are difficult to assess precisely, nevertheless workers appear to be aware of many important health-related concerns.

Ideally, one might like to compare workers' job risk perceptions with an objective index of the hazard actually faced to ascertain the extent to which these judgments are accurate. Such a comparison is well beyond the capacity of existing data and in some cases beyond our limited understanding of the determinants of the "true" risk levels of a job. However, one can compare the risk perceptions of workers in various industries of different overall risk levels. The principal drawback of this procedure is that the overall industry record may not accurately characterize the risk of each job within the industry. Despite this limitation, the evidence in Table 4.2 is quite striking. The first column distinguishes different levels of industry-wide risk, where the hazard measure is the injury frequency rate for the worker's industry.[2] Most of the workers sampled were in relatively safe industries, as indicated by the distribution in the second column. As shown in the final column, the fraction of workers in each risk group who consider their job hazardous increases fairly steadily with the industry injury rate.[3] Indeed, all workers in the most hazardous industry group consider their jobs hazardous.

Although there is no way of ascertaining whether the workers' beliefs coincide with actual risk levels, there does appear to be widespread worker perception of both health and safety risks. Moreover, these perceptions appear to be reasonably well founded, since they closely parallel objective job risk measures.

LEARNING ABOUT JOB RISKS AND THE MARKET RESPONSE

These risk perception data do not indicate when a worker begins to perceive his job as hazardous.[4] In the idealized compensating

TABLE 4.2. RELATION OF JOB RISK PERCEPTIONS TO INDUSTRY RISK LEVELS

Industry injury rate (IR)	Fraction of workers in sample	Fraction in group who view job as dangerous
$0 \leq \text{IR} < 5$	.50	.24
$5 \leq \text{IR} < 10$	.18	.43
$10 \leq \text{IR} < 15$	.08	.47
$15 \leq \text{IR} < 20$	.08	.53
$20 \leq \text{IR} < 25$	.06	.68
$25 \leq \text{IR} < 30$	.07	.66
$30 \leq \text{IR} < 35$	.01	.64
$35 \leq \text{IR} < 40$	.02	.60
$40 \leq \text{IR}$	.01	1.00

Source: Based on Viscusi (1979a), Table 14.8.

differential model, workers' risk perceptions are always accurate and consequently do not change after beginning work on the job. In the usual situation, work experience enables the worker to reassess the risk he faces and to quit if he finds the risk unattractive.[5] This potential for learning gives rise to a job hazard–quit relationship, which is the focus of this section.

In the typical learning process, the worker uses different sources of information to continually update his assessment of the job risk.[6] Many workplace characteristics that generate risks or are correlated with job hazards can be readily observed—safety training procedures, noise level, the presence of noxious fumes, and the degree of care taken by coworkers. Injuries to himself and others also may be instructive in forming a worker's risk assessments.

The impact of this learning on worker decisions hinges on being able to acquire some information about the risk and to use this information when facing similar risks in the future. Learning of this type is widespread. The highly publicized problems at the Three Mile Island nuclear reactor, for example, clearly led to a major change in the public's perception of nuclear safety. If we exclude the role of anxiety about risks, this change in perceptions is only relevant to our welfare insofar as future decisions will be affected by it. Since individual decisions to live near nuclear plants, as well as future nuclear safety policies, undoubtedly will be altered by the shift in risk perceptions induced by this accident, learning about nuclear risks has an economic value. The high incidence of lung cancer among shipyard workers with previous exposure to asbestos has had a similarly dramatic effect on our risk perceptions, but this information primarily benefits new workers rather than those who have suffered irreversible damage from exposure over the past few decades.

For many chronic diseases, it may be difficult for a worker to make any reliable causal inferences that would improve future job decisions. One cannot be confident of the efficiency of this learning process until one first ascertains whether learning will occur and, if it does, whether it will enable those now incurring the hazards to make sounder job decisions. It is possible, however, to make some overall judgments, using the information we have on worker perceptions.

Statistical analyses of the determinants of risk perceptions cannot be conclusive, but they are consistent with the relationships that would hold if workers formed their risk assessments before working on the job and then revised these assessments based on subsequent information. As indicated in Table 4.2, workers in more hazardous industries are more likely to view their jobs as hazardous, as one would expect if the general reputation of the industry influenced their initial risk perceptions. Characteristics of the job, such as the physical effort required, the regularity of work, and the pleasantness of the physical conditions, are also instrumental in affecting workers' assessment of risks.

The most powerful determinant of a worker's risk perception is his own injury experience. Seventy-one percent of all workers who have experienced a job-related injury or illness view their job as hazardous (see Table 4.3). Taking into account other determinants of risk perceptions, including the overall riskiness of the industry, a worker's being injured increases the probability that he will view his job as hazardous by 0.2.

It is particularly striking that almost one-third of all workers who had experienced injuries did not view their jobs as hazardous. This aberrational result appears to be due mainly to illnesses and injuries that occurred on the job but were not an intrinsic part of it. The injuries least likely to be associated with

TABLE 4.3. DANGER PERCEPTIONS AND INJURY EXPERIENCE BREAKDOWN FOR BLUE-COLLAR WORKERS

Worker category	Proportion who consider job hazardous	Proportion who do not consider job hazardous
Injured on current job	0.10	0.04
Not injured on current job	0.42	0.44

Source: Viscusi (1979a), Table 14.9.

risk perceptions were: muscle or joint inflammation, poisoning, dermatitis, and other job-aggravated ailments. Since these injuries are principally due to the worker's own limitations (job-aggravated ailments) or unusual acts of carelessness (poisoning), it is not surprising that many injured workers do not view their jobs as the source of the hazard.

Overall, workers' risk perceptions accord with what one would expect if learning about risks occurred on the job. A more conclusive test for this learning effect would be possible if we had evidence on the evolution of workers' risk perceptions and the information that influenced these beliefs. The data we do have enable us to conclude only that those who work in unpleasant environments or who have been injured are more likely to view their jobs as risky.[7]

The empirical implications of this learning process are more clear-cut. I will view workers as being engaged in the following adaptive process. A worker's decision to accept a potentially hazardous job is based on his assessment of the risk, the wage rate paid, and his expectations regarding the implications of the job for his future well-being. Risk premiums are still relevant since increases in the assessed probabilities of injury boost the wage that the worker will require to take the job. If the worker's on-the-job experience generates sufficiently adverse risk information, given this wage rate, he will quit.

We would observe no job hazard–quit relationship if workers fully understood the risks before starting the job. Similarly, if workers never learned about the risks they face, they would remain with the position until they retired, became disabled, or switched jobs because of other economic factors. In either case there would be no relationship between hazards and quitting.

This adaptive behavior is economically important partly because of the somewhat surprising aspects of the relationship, most notably workers' predilection for jobs posing risks that are not well understood. Before exploring these properties, it is useful to obtain some perspective on the empirical significance of this phenomenon. If few workers do quit their jobs after learning about the risk, the more subtle aspects of the analysis might best be regarded as theoretical curiosities.

To isolate the independent effect of injury rates on quitting, one should control for other factors that may induce workers to quit. This is especially important for variables correlated with injury rates. Using a statistical model that included the industry wage rate, unionization, worker mix variables (age, race, and sex mix), and a variety of other industry characteristics, I distinguished the effect of injury rates on aggregative quit rates from these other influences. The impact is quite dramatic, as worker injury and illness rates may account for as much as one-third of all quits in manufacturing industries. Some of these quits may be due not to learning per se but to a change in the worker's physical capabilities after becoming injured or disabled. To resolve this issue, we must utilize data on individual quit behavior, taking into account the individual's health status.

Using the blue-collar worker data discussed earlier, I found that workers' risk perceptions had a powerful influence on their intentions to quit. The chance that the worker claimed that he was very likely to look for a new job with another employer in the next year was increased from 0.10 to 0.21—more than double—if the worker viewed his job as hazardous in some respect. Hazard perceptions had an identical effect on more moderate quit intentions (whether it was somewhat likely that the worker would search for an alternative job). Industry injury rates have a similar impact on job satisfaction and job search activity in other sets of survey data.

Without evidence on workers' job changes, it is not possible to assess whether workers' intentions to quit are fulfilled. Using a sample of almost 6,000 workers, I have estimated that industry injury and illness rates boost workers' propensities to quit by 35 percent.[8] This estimate is comparable to the effect of job risks on quit behavior shown by using aggregate manufacturing industry data.

The job risk–quit phenomenon takes on an additional dimension when viewed over a longer term. The average total length of employment at a hazardous firm tends to be less if workers quit more often, so more hazardous enterprises tend to have less experienced workforces. Inexperienced workers will tend to have more injuries because there will be greater concentrations of these employees in high-risk situations. Even if workers are not more accident-prone when they start the job, we would observe a strong relationship between job risks and the number of years of workers' experience at the firm.

This pattern is reinforced by firms' assignments of workers to different jobs. Since firms have a substantial investment in the training of experienced workers, they attempt to hold down these costs by assigning inexperienced workers to hazardous positions, as in the case of the B. F. Goodrich Company, where the polyvinyl chloride exposures were greatest for those in entry-level positions.

The oft-cited statistics on the high accident rates of inexperienced workers consequently overstate the extent to which workers are responsible for the accidents. Less experienced workers tend to be more accident-prone because the employer assigns them to high-risk jobs and because the high rate of quitting from these positions further reduces the average experience of the group. The conventional attribution of accidents to irresponsible behavior of inexperienced workers may be in large part misplaced.

The most fundamental implication of the job hazard–quit process pertains to the nature of workers' job choices. In particular, when choosing among jobs posing comparable initial risks, should the worker select the job posing a precisely known risk or a job posing a risk about which little is known? He will tend to prefer the more uncertain job since it affords the opportunity for on-the-job learning about the risk. If the worker finds out that the job is very risky he can quit, and if he learns that it is safe he can remain on the job.

Even in situations in which quitting is not feasible, as when the only on-the-job learning consists of finding out whether or not he becomes permanently disabled, will the worker prefer the uncertain job? Perhaps somewhat paradoxically, a worker in this situation will still prefer the job risk that is only dimly understood since it offers a greater chance for long-term survival. If both jobs offer the same initial chance of survival, a series of successful outcomes on the uncertain job will lead the worker to revise downward his assessment of the risk, making that job more attractive. If, however, the worker becomes permanently disabled and must leave his job, his subsequent upward revision of the risk assessment is irrelevant since it will not alter his decision to remain on the uncertain job. The only information pertinent to future decisions is favorable. With a sequence of job risk lotteries that terminates with an unfavorable outcome, such as those on life and death, workers will prefer the uncertain job, for any given initial level of risk, since it offers the greatest chance of a continuous streak of favorable outcomes. The net effect is to create a bias in workers' decisions toward jobs posing risks that are not well understood (see Appendix). Before assessing the overall implications of this behavior, let us first consider possible responses by their employer in this situation.

Imperfect Information and Enterprise Decisions

In most firms worker turnover is costly because of the hiring and training costs involved whenever a worker is replaced. The firm could reduce these costs by decreasing its training investment through job assignment and training practices. Alternatively, it could reduce costs by eliminating the job risk–quit relationship, which the firm can do in a variety of ways. First, it can attempt to attract only these employees who will not quit. Second, it can adopt a technology whose level of risk is well known. Finally, it can provide complete job risk information to workers to eliminate any subsequent learning. I will consider whether it is in the employer's interest to take these actions and, if so, whether learning-induced quits will be eliminated.

If workers knew before accepting the job that they would quit because of the risks, the employer could eliminate these turnover costs quite readily without having to identify the quit-prone workers. In particular, it could pay employees a very low salary during their initial years with the firm and guarantee a very high wage if they remain for a substantial period. Workers who expect to quit will not find this wage structure attractive since they do not expect to remain with the firm long enough to reap the high wage. Tilting the wage structure in this fashion will lead the workers themselves to choose whether they will work at the firm, thus screening out the quit-prone employees.

In the case of learning-induced quits, this wage policy cannot be fully effective since the workers do not know whether they will quit until after some period of work on the job.[9] Furthermore, if turnover costs are small, it will be in the employer's financial interest to attract imperfectly informed workers to the firm. Workers' preference for jobs with less well-known risks enables firms to offer a lower wage than they would otherwise. Since the wage bill increases as workers' probabilistic beliefs become more precise, employers have a financial interest in hiring workers who are not cognizant of the risks they will face.

Very high turnover costs might make it desirable for a firm to prevent quitting by raising the wage rate. However, the level of turnover costs is not predetermined but is typically subject to the discretion of the firm. If the job risks in a firm generate quit behavior, the firm will reduce its hiring and training expenditures, since worker turnover diminishes the value of this investment. Actual training patterns reflect this influence, as workers in high-risk industries are less likely to receive any formal job training.[10] Only in a highly unusual situation would we expect to see a firm sacrifice the wage reductions that are possible if it does not attempt to eliminate learning-induced quits.

Similarly, there is little financial incentive for a firm to reduce hazard-related quits by adopting a technology whose risks are well known. From the standpoint of the wages it must pay, the firm will find the uncertain technology preferable. Under the relatively neutral assumption that workers' initial risk judgments are unbiased, the firm will adopt a technology that is not well understood and that poses a higher risk than would be optimal if workers fully understood the risk.[11]

The most prominent occupational health risks usually involve relatively new technologies whose properties are not yet known. The problem may be a newly developed chemical that is potentially carcinogenic, or it may be longer-term occupational exposure to substances such as asbestos whose implications have been publicized only in the past decade. In either case, those formulating policy are often quite reluctant to impose restrictions that will discourage attractive new technologies or existing technologies with uncertain properties.

Whether or not these concerns are justified can be assessed only after first considering the nature of technological choice in the market. Since market forces will tend to be biased toward overly risky technologies that are not well understood, the economic benefits from technological progress should be balanced against the adverse effects stemming from technologies posing inefficiently large risks. An alternative to choosing technologies with widely known riskiness is to provide risk information to workers so that they can assess these hazards more precisely. Available evidence suggests that few firms make a comprehensive effort to inform workers of the risks they face. For example, no firms tell their employees the average annual death risk they face.

Much information that firms do provide is not intended to enable workers to assess the risk more accurately. Rather, it is directed at lowering workers' assessment of the risk. The most widespread claim by firms is that National Safety Council statistics indicate that the worker is safer at work than at home—a statement that, as I mentioned earlier, is intentionally misleading.

The kinds of risk information given to workers covered by collective bargaining agreements is summarized in Table 4.4. Most of the limited information provided is not intended simply to inform workers of the risks; the principal purpose is to decrease accidents through various safety promotion activities. Twenty-seven percent of unionized workers are covered by agreements in which the employer provides employees with safety information, such as safety rules or warning signs. In a slightly greater number of instances the union itself receives this information. Only 15 percent of the agreements call for formal provision of accident, mortality, and morbidity data to the union.

The most ambitious case of risk information is the paint and

TABLE 4.4. INFORMATION TRANSFER PROVISIONS IN COLLECTIVE
BARGAINING AGREEMENTS, 1974–75

Type of information transfer	Agreements with such provisions	Workers covered (thousands)
Dissemination of safety information to employees	273	2,130
Safety rules and procedures	186	992
Possible job hazards	73	1,122
Warning signs or labels	74	683
Dissemination of safety information to union	335	2,165
Safety rules and changes	58	405
Reports, minutes of safety meetings	158	1,477
Accident, mortality, morbidity data	213	1,210
Safety education and training	132	1,497
Safety inspections	335	2,357
Union-management cooperation on safety	757	3,946
Total agreements studied	1,724	7,868

Source: U.S. Bureau of Labor Statistics (1976), pp. 4–55.

coating industry's chemical labeling system. Labels on containers indicate the level and nature of the hazard, the need for particular kinds of protective equipment, and the need to exercise special kinds of care (no smoking near flammable substances). Although this newly instituted system will affect workers' risk perceptions, its intent is to reduce accidents and the companies' premiums for workers' compensation.

Information directed at improving safety, such as safety training procedures, has two types of effects. First, it may alter workplace safety levels. As such, it can be analyzed in much the same manner as other enterprise investments in the technology.

Second, providing safety information has purely informational aspects, which will be the focus of the discussion here.

Providing workers with additional information about job risks will sharpen their assessments of the risks, which will tend to increase the wages the firm must pay. Unless the information also lowers the assessed probabilities of being killed or injured, it will not be in the employer's interest to provide it unless the costs of worker turnover are high. The same types of influences that discourage screening out workers who are likely to quit or utilizing technologies with well-known properties also inhibit the provision of information.

Since workers could improve their job choices if they knew of the risks posed by different positions, one might expect some market mechanism to emerge that would enable workers to purchase job risk information from their employer. Such arrangements have not evolved, owing to the aberrant characteristics of information as an economic good.[12] Since information is costly to provide, particularly in a form that can be readily processed by workers, firms' failure to inform workers of the risks they face is not sufficient to prove an inadequacy in the way markets function. Instead, one must assess the importance of the economic factors limiting information provision.

Unlike commodities in usual market transactions, the use of this good, information, is not restricted to the purchaser. If, for example, a firm reveals the job risks to its current employees, it will be difficult to deny prospective workers access to this knowledge. At best, the employer could charge its workers for information and extract from them the total amount they would be willing to pay for it. The firm could not recoup any damage to its reputation or any increases in future wage rates that resulted from adverse information. Similar difficulties may inhibit efforts to provide information to workers who are matched to inappropriate jobs within a firm. If the risk perceptions of other workers are affected adversely, it may not be desirable to provide information that would reduce accidents by improving job-worker matchups.

The economic properties of information are tainted in another way by the employer's obvious direct interest in its use. An enterprise will be willing to disclose very adverse information only at a very high price. Similarly, information provided cheaply to workers must be relatively favorable, since otherwise the firm will be forced to pay higher wages. If the worker knows he can learn about the risks of his job for some specific cost, such as fifty dollars, he will know by the cheap price that the job is relatively safe.

A more general fundamental problem with selling information is that the buyer cannot know how valuable the information will be to him until after it is revealed.[13] A company might be willing to disclose a new carcinogenic risk if it could collect the total value that workers would have placed, in retrospect, on this information. However, workers would offer very little for this information if their prior assessed probabilities for a possible risk of this type are close to zero.

A final difficulty depends on how workers' risk preferences are formed. Suppose that a worker bases his judgment on the average risk of a particular occupation across the entire industry. This approach is not entirely unreasonable if the positions are fairly uniform, since it enlarges the size of the sample that can be used in assessing the risk. A firm's specific performance will affect the worker's judgments only by altering the perception of the average industry risk. In this case, the incentives for the firm to provide risk information or a safe work environment will be diluted, since it cannot recoup any of the benefits of increased safety that it confers on the rest of the industry.

Since it is typically not in the firm's self-interest to provide job risk information to resolve worker uncertainties, job choices will be made on the basis of imperfect information. Incomplete knowledge of the risks does not simply erode the efficacy of the compensating differential process. In many instances, workers have the opportunity to learn about these risks, and if they choose not to incur them, they will quit. The resulting job hazard–quit relationship is borne out quite strongly in available data. Although the relative importance of risk premiums and learning-induced quits cannot be meaningfully compared, each of these mechanisms should be viewed as major parts of an inter-related process by which workers attempt to match themselves to appropriate jobs. These market processes do not, however, ensure that outcomes will be fully efficient. With imperfect information, workers facing risks on a continuing basis will display a systematic preference for risks that are not precisely known.

To assess the implications for market behavior, we must make some assumptions about workers' initial risk assessments. The most neutral assumption is that their initial judgments are unbiased. If this is the case, we can draw three principal conclusions.[14] First, the level of risk provided in the market will be higher than is optimal. Second, there will be a bias toward new technologies that pose risks that are not well understood. Finally, the level of insurance compensation for injured workers will be set at too low a level in market agreements between employers and workers. Each of these effects can serve as the basis for government intervention.

5

THE BASIS FOR GOVERNMENT INTERVENTION

ALTHOUGH JOB RISKS ARE THE OUTCOME OF SYSTEMATIC LABOR market processes, deficiencies in these mechanisms may create a potential role for governmental policy. In this chapter I will review the inadequacies of the market, the criteria for designing efficient regulatory policies, and alternative modes of intervention. Subsequent chapters will address the implications of these guidelines for the design of risk regulations.

The form of market failure that has dominated my discussion thus far and that is most salient in analyses of risk regulation is the inadequacy of information about the risks being incurred. If workers and firms are not fully cognizant of the job risks resulting from their decisions, the desirable properties usually imputed to market outcomes may not prevail.

Two types of informational inadequacies can be distinguished. First, workers (and firms) may have biased perceptions of the risk, that is, they may systematically overestimate or underestimate the hazards involved. The usual claim is that workers tend to underestimate risks. In the rare cases where these assertions are supported, it is by citing evidence based on individual probability assessments in test situations in which respondents facing hypothetical decisions tend to underestimate small probabilities.[1] The relevance of these experimental findings is unclear. Participants in such experiments have less of an incentive to think systematically about small probabilities than do workers whose lives may hinge on whether they know the risks. Moreover, for the average worker the chance of dying can justifiably be called rare (with an annual risk of 1/10,000), but the risk of injury and illness is quite substantial (with an annual risk of over 1/30). The widespread evidence that the market does provide risk premiums suggests that workers are quite aware of job hazards. Even more compelling is the evidence on worker perceptions of risk; subjective and objective risk measures are strongly correlated and lead to virtually identical estimates of job hazard premiums.

These findings do not foreclose the possibility that workers' assessments of job risks are biased, and they do not indicate the direction of any systematic bias that may exist. Until we obtain more refined data on actual and perceived job risks, it will not be possible to resolve these issues. At present, there appears to be little support for a general presumption that workers typically underestimate risks, but this possibility should be considered within the context of the specific type of hazard.

The second form of imperfect information, which stems from the imprecision of worker judgments, was the focus of the previous chapter. This form is associated with a general behavioral pattern—hazard-induced quitting—that is consistent with worker actions and, unlike assumptions about biased perceptions, does not rest on strong assumptions about market behavior. Even if workers do not make biased risk judgments initially, they will systematically prefer jobs whose risks are not fully understood. While on-the-job learning about risks and subsequent quit behavior serves as a beneficial market response, the quit mechanism does not guarantee efficient outcomes. In the extreme case in which a worker learns that he has contracted a fatal disease, for example, his welfare has been affected irreversibly, so quitting will do him no good. Under a broad set of circumstances, the imprecision in workers' job risk beliefs will lead to too high a level of risk, a bias toward technologies that are not fully understood, and too little insurance coverage for workers in hazardous jobs.

The extent to which lack of risk information poses significant problems for worker decisions depends on the type of risk.

Worker information about external safety risks (such as the risk of falling off a platform) is usually greater than for internal health risks (such as the effects of radiation). First, the factors contributing to safety risks, such as a loose handrail, are often readily visible, so the worker can use them in formulating risk judgments. Second, accident outcomes can be readily observed, monitored, and publicized. And finally, most safety hazards cause accidents with sufficient frequency that there is a large body of evidence available for formulating one's judgments.

In contrast, many health hazards pose problems in each of these respects. Although radiation levels and some noxious fumes are identifiable, in many cases the nature and implications of the work environment are not well understood by anyone, including the worker. Adverse health outcomes, such as cancer or hearing loss, are not recorded with the same accuracy as accidents, and even when the ailment is known, the cause may be unclear. Is the lung cancer of a cigarette-smoking foundry worker due to his smoking, his current job, or previous activities? The long gestation period—a decade or more for many types of cancer—impedes attempts to infer causal linkages of this type. Moreover, the exceedingly low probabilities of many health risks greatly diminish the opportunity to distinguish the relative impact of these diverse influences.

Not all health risks are underassessed, however; newly discovered and widely publicized risks may be overassessed. More generally, since 40 percent of all risks cited by workers are health-related, one should be careful not to assume that the greater difficulties in assessing health risks means there is no knowledge about these hazards.

A second potential basis for government regulation of job risks stems from inadequate insurance for such risks. Even if insurance markets function effectively, the level of insurance provided will not be ideal if workers do not accurately perceive the risks that are to be insured. The available insurance arrangements also may be inadequate. Problems of adverse selection (only the bad risks join, threatening the viability of the plan) affect most types of insurance to some degree and serve as a rationale for a mandatory workers' compensation system. Under a voluntary workers' compensation system, workers in high-risk jobs would be most likely to join, raising the average premium charged and consequently discouraging workers in lower-risk jobs from joining. As the pool of covered workers became increasingly risky, the fraction of the workforce insured by the program would steadily decline. The critical deficiency of insurance markets is quite fundamental and unalterable; there is no market at all for what is at risk—life, hearing, and individual health. Since these fundamental determinants of welfare cannot be transferred, no workers' compensation or insurance system can ever fully ameliorate the inadequacies of insurance market outcomes.

A third possible justification for intervention is the concern of society at large for those whose health is adversely affected by work. These concerns may be financial or strictly altruistic. Taxpayers have a direct financial stake in worker health since injured workers boost the costs of a variety of social insurance programs, including welfare and aid to the disabled. To the extent that the purpose of risk regulations is to reduce the costs of social insurance programs, we should ascertain whether these efforts are well designed, to avoid the problems associated with pyramiding government intervention.[2] Otherwise, an ill-conceived social insurance program may provide the justification for additional government intervention designed to reduce its costs. Society would be better off without either program.

Altruistic concerns with individual health are pervasive and are reflected in society's decision to promote medical research and to provide subsidized medical insurance to the aged and poor. In some circumstances, these concerns may be reflected in market decisions. For example, a worker may take into account his family's altruistic concern with his welfare when making his job choice, but these concerns and those of society at large may not be fully reflected in his decision.

While altruism may be a legitimate basis for intervention, it raises two relatively intractable problems. First, since society's concerns are not transmitted through market forces, we have no way to measure magnitude of the concern. As one might expect, workers place a considerable value on their own well-being (see Chapter 6). It is unclear whether recognition of the value that others place on the worker's health would significantly alter the overall value we attach to his well-being.

Even more problematic is that altruistic concerns typically are intermingled with wealth-related differences in preferences. Since risky jobs become more unattractive the higher one's income class, we should expect that many of those who are most outspoken on policy issues will consider the jobs of workers in hazardous industries abhorrent. While perhaps well intended, intervention based solely on these grounds will necessarily reduce the welfare of the poorer workers in society, as perceived by them.

A more recent variant of the altruistic arguments asserts that individuals have a right to a safe work environment.[3] Efforts to promote present risk regulations on the basis that they enhance worker rights are certainly misguided. Uniform standards do not enlarge worker choices; they deprive workers of the opportunity to select the job most appropriate to their own risk preferences. The actual "rights" issue involved is whether those in upper income groups have a right to impose their job risk preferences on the poor. A superior method of expanding worker rights is to provide the risk information that will enable workers to make choices that better reflect their individual preferences.

Any inadequacy in market processes, including many less prominently discussed forms of market failure,[4] could create a role for government regulation of risks. But even rampant instances of market failure do not ensure that any of the available government policies will be beneficial. The policy problem is to design and implement programs that do not undermine the constructive aspects of an imperfect market but that enhance the general welfare of society.

CRITERIA FOR POLICY EVALUATION

The dominant principles for selecting the appropriate policy are closely related efficiency criteria. First, the government should select the policy that provides the greatest excess of benefits over its costs and, since one alternative is to do nothing, it should not adopt any policy whose costs exceed its benefits. Second, to obtain the highest net gains from policies, the scale of the programs should be set at levels where the incremental benefits just equal the incremental costs; further expansion or reduction in the policy will produce lower net benefits overall.[5] Third, all policies should be cost-effective, that is, the cost imposed per unit of benefit should not be greater than for other policies. All policies meeting the first benefit-cost criterion above are necessarily cost-effective. Cost-effectiveness analysis is useful in situations in which it is difficult to ascertain the worth of the particular benefit, such as a certain number of cases of heart disease prevented, but in which there is a desire to provide these benefits as efficiently as possible. Policies may be cost-effective but still undesirable, since there is no assurance that the level of benefits is optimal, only that the level is provided at the lowest possible cost.

Although these three policy evaluation criteria have considerable intuitive appeal, they have not gone unchallenged. Two types of criticisms can be levied: first, the principles may not be utilized correctly, and second, they may not be compelling even under ideal circumstances. The first class of objections, which dominates most discussions, consists largely of problems of implementation. The evaluation may not include all categories of benefits and costs, some categories of effects may be evaluated more accurately than others, and important policy alternatives may be ignored. For example, the four decades of government experience with benefit-cost analysis of water resources projects has been characterized by flagrant underestimation of project costs, inflated benefit estimates, and selection of projects on the basis of narrow, pork-barrel interests.[6] While there could be similar abuses in the design of risk regulation policies, this does not imply that a systematic assessment of the merits of policies necessarily produces inferior policy choices.

The benefit-cost guidelines themselves are not intrinsically flawed. Rather, particular evaluations of policies may place undue emphasis on certain effects or ignore relevant concerns. Unless thorough policy analyses systematically enhance the influence of policymakers who are incompetent or who promote policies in conflict with the objectives of society at large, it is doubtful that one can even make a strong pragmatic argument that ignorance of the impacts of policies is a preferable basis for government decisions. Certainly the performance of the more intuitive school of policymaking, as characterized by the past operations of all risk regulation agencies, has shown that the disregard of broad classes of policy effects—notably their costs—does not further society's objectives.

The second and more fundamental criticism of policies based on efficiency criteria merits more serious consideration. While a comprehensive tally of benefits and costs appears unobjectionable, it may raise quite legitimate distributional issues. Since the benefits and costs accrue to different groups, the process of aggregating these effects abstracts from the distribution of the impacts, as all gains and losses are treated symmetrically. The justification for such aggregation is that if a policy's benefits exceed its costs, those who benefit can potentially compensate those made worse off, so some groups can be unambiguously better off while the welfare of others is unchanged.[7] However, if the compensation is not actually paid (it seldom is paid), there will be winners and losers from the policies. The judgment that these effects should be treated symmetrically in this situation, which is implicit in the benefit-cost calculus, is a legitimate object of controversy, though it might not be an unreasonable basis for designing most policies.

The case of such distributional conflicts for social policy in general is particularly acute when productivity is not growing, since any redistribution of income will then result in gains for some groups, offset by equal losses for others. The sum of these effects is zero, giving rise to what Thurow has termed the "zero-sum society."[8] Many observers have extrapolated this zero-sum notion to the case of occupational health and safety policies, with gains to workers from increased regulation being matched by equivalent losses to firms.

This concept has little applicability to risk regulations, however, since these policies rarely offer zero-sum choices concerning who should benefit most. Rather, ill-conceived policies are typically negative-sum, since the losses are in excess of the gains. Overly stringent risk regulations may impose costs on firms and may deprive workers of jobs they would have wanted. All parties involved may be made worse off if policies are misdirected.

Similarly, policies with benefits in excess of their costs are necessarily positive-sum. For most policies introduced in unregulated market situations, the gains to workers are greater than any gains to firms. Firms also could benefit from these policies if, for example, workers believed that their jobs were safer, leading them to accept lower wages. The fact that few firms are clamoring for additional government intervention suggests it is doubtful that present risk regulation policies are beneficial to the business community. The central difficulty raised by the divergence of interests is that many observers interpret it as evidence that the only issue at stake is the division of society's resources. While this is an obviously important issue, the total value of society's resources will also be affected by these initiatives.

A recurring distributional concern is whether the benefits should be tilted toward a particular income group. Policies to promote workers' health and safety do not enhance the well-being of those who have no job at all. Those who might become unemployed as a result of these policies also seldom benefit. If very stringent and inefficient policies are adopted, many hazardous jobs that individuals were willing to take will be eliminated. Less affluent workers will suffer the most since they have the greatest willingness to boost their income through potentially risky jobs. Whether this interference with individual choices is warranted depends on the justification and manner of intervention and, most important, whether these regulations have arisen simply from the risk preferences of those responsible for the policies, without consideration of the workers' risk preferences.

Even in situations where there are legitimate distributional concerns, policies that promote income redistribution should be designed in the most efficient manner. Since society has a variety of policy tools to assist the poor, such as welfare benefits and subsidized health insurance, risk regulation policies should not be distorted to promote distributional objectives if these other policies are better suited to this purpose.

The workers' compensation program is similar to these social insurance efforts and also can be used to aid the poor without causing serious distortions. Benefits to the disabled poor could be set at relatively higher levels since they have fewer additional resources to meet their needs. Some tilting of the benefit structure already occurs, since the various limits on benefits lead to higher rates of income replacement for low-wage workers.

The most prevalent form of distributional preference actually arises implicitly in the policy design process. Regulatory analyses typically treat the benefits of risk reduction identically for all income classes of beneficiaries. Since affluent workers are much more willing to pay for these benefits, the use of some average valuation for all individuals in effect overstates the benefits to the poorer and understates the benefits to the richer. This form of implicit redistribution may remain the most prevalent method of making distributional distinctions, because it arises from symmetric treatment of individuals in situations in which it may be quite difficult to make explicit distinctions.

PROVISION OF INFORMATION

Designing policy to control risks involves not only assessment of policies, but also development of a viable set of alternatives from which to choose. Chief among these alternatives are the provision of risk information so that workers and firms can make more knowledgeable decisions, compensation of workers after they are injured, and direct regulation of workplace conditions either through standards or penalties linked to adverse working conditions or job outcomes. A final policy option is to have no government intervention and to rely on market forces.

The least obtrusive form of intervention is provision of information. One reason for the government to take an active role in this area is the failure of the market to provide the kind of job risk information the worker needs to make informed choices.

The most traditional government policy role has been the collection and dissemination of injury and illness rate statistics, which make it possible for analysts to form more reliable judgments on risk trends overall and in specific industries. The pooling of risk information across firms also could potentially enable policymakers to make inferences about the determinants of risk that would not be apparent to those at the firm. For example, relatively rare forms of cancer may be dismissed as unrepresentative random events when encountered in an individual firm, but on an industry-wide basis may be seen as part of a broader pattern of illness related to some new carcinogenic exposure.

Unfortunately, very little use has been made of the capacity offered by a centralized system of risk information. Not only has there been minimal risk-related research with these data, but there is no longer an opportunity to undertake such studies, since OSHA no longer has access to the injury and illness data. In a classic case of bureaucratic imperialism, the Bureau of Labor Statistics now refuses to give any of its firm-specific data to OSHA or to any other group outside BLS. Only the published industry-wide averages for broad risk categories are available. Resolution of this intradepartmental conflict is a prerequisite for any meaningful policy of risk information and, more generally, for OSHA's policy evaluation effort.

Using this risk information, OSHA could educate workers about relative risks much more extensively than under its present requirements for accident log books. Ideally, an information policy should enable workers to make more informed choices and to select the wage-risk combination they prefer. When there is substantial heterogeneity in the society in attitudes toward risk, more stringent regulatory policies that limit the range of job choices sacrifice a fundamental benefit of the market system.

The provision of information could take a variety of forms. Workers could be apprised of the injury and illness rate for the firm, its death rate, or the accident record for a particular job. Another possibility is to introduce a qualitative grading system for jobs based on relative severities of risk. Employers also could be required to discuss with workers the impact of their actions on the risk level, such as the effect of cigarette smoking on the risks from asbestos exposures and the need to exercise special care in different work situations. If workers are overwhelmed with risk information, however, so that they are unable to effectively utilize it in making their decisions, their choices might not be improved.

An innovative form of information provision would be to identify the relative risk of the various firms in the industry.[9] When workers judge their job risk by the average industry risk, the information which is most readily available, relatively safe firms are penalized and have less incentive to invest in workplace safety. If the government identified the safest firms, it would reward enterprises that invested in safety and would also lead other firms to raise their safety levels to avoid being lumped with the pool of unsafe firms. In effect, identifying safe firms establishes an unraveling process, as more and more firms attempt not only to reap the benefits of being identified as safe but also seek to avoid the costs associated with being in the unsafe group. As additional firms leave the unsafe classification, the average riskiness of the group rises, and the cost of being in it steadily increases.

To the extent that firms' risk records have been publicized, it has taken the form of identifying the most hazardous enterprises where major catastrophes have occurred or where particularly unusual illnesses have been identified. This practice may provide safety incentives to the particular firm receiving the adverse publicity, but it does little to provide safety incentives to other firms, which may even be adversely affected despite safe workplace conditions if the risks of the most hazardous firms are presumed to be industry-wide.

Business groups are usually more receptive to identification of the highest-quality rather than the lowest-quality firms. In some cases, relatively safe enterprises, such as DuPont, have made significant efforts to make their good performance known. Such efforts have a quite legitimate economic justification and could serve as the basis for a more active governmental role in identifying safer firms.

Although several types of risk information policies appear promising, the question of how much or what kind of information will prove useful cannot be resolved without analyzing how workers actually respond to it. Rather than restrict the policy to providing a single type of information, it would be preferable to experiment with several alternative forms and to analyze the effect on worker wages, enterprise investments in safety, accident and illness rates, and worker turnover. I include more explicit suggestions along these lines in my policy proposals in the final chapter.

In addition to processing and disseminating injury and illness data, the government can also serve a productive function by undertaking basic research into the health implications of different working conditions. The National Institute of Occupational Safety and Health (NIOSH), a division of the Department of Health and Human Services, is responsible for this research effort; its focus has been on ascertaining the properties of potentially carcinogenic substances. The limitations of this research are largely determined by the nature of the policies for which the research is undertaken. The present focus of policy on risk thresholds and potential carcinogenicity is shared by the scientific research. Once policies are based on the overall desirability of regulation rather than on the level at which a risk can be identified, the incentive to provide relevant information will be enhanced.

COMPENSATION OF WORKERS

In addition to providing risk information to workers and firms, the government could attempt to regulate the risk itself by preventing worker injuries and illnesses, or it could compensate workers after the health effects have occurred.[10] These compensation schemes may affect the risk level through their financing mechanisms.

The use of some type of *ex post* compensation scheme is quite common in risk regulation. The Environmental Protection

Agency "Superfund" policy, for example, addresses the toxic waste problem through a fund used to clean up hazardous waste sites. One of the justifications for a central clean-up fund is that the firm responsible for the waste often cannot be identified, so liability is difficult to assign. In the case of worker injuries, there is typically no problem of identifying the parties involved, but the liability of these parties is often unclear since irresponsible worker actions and activities (smoking or careless behavior) may influence the outcome observed. Sorting out the respective responsibilities of the parties would be quite costly and a highly imperfect process.

Establishment of a no-fault system of compensation eliminated the costs associated with assigning responsibility for accidents. The extent to which workers have become less responsible about safety is limited in part by the continued undesirability of being injured, since workers' compensation benefits do not eliminate all losses in welfare. Perhaps more important is that employers can continue to monitor workers' safety-related behavior, penalizing or firing workers who do not exercise care.

This ability to monitor and reward worker behavior is a principal reason why a no-fault workers' compensation system is viable, whereas a no-fault product liability system has not emerged. The manner in which consumer products are used cannot be monitored or controlled, and in some cases one cannot even ascertain whether a particular product was in use at the time of the accident. Advocates of extending the workers' compensation no-fault principle to other risk contexts typically ignore the importance of being able to substitute employer control of worker behavior for the economic incentives to behave safely that are sacrificed under a no-fault scheme.

State systems of workmen's compensation were established not only to eliminate the costs of ascertaining responsibility for accidents but also to provide income support and medical assistance to injured workers. The objective of workers' compensation programs is not to eliminate all losses in welfare resulting from job hazards. If a worker is killed, for example, no financial reward to his family can compensate him for his death. Even if he knew before his death that his family would receive such a bequest, considerable transfers of funds are unlikely to make him indifferent to dying. The central difficulty is that we can transfer money, goods, and services, but we cannot transfer lives and the fundamental determinants of physical well-being.

A key policy question is how much income support the system should provide. An instructive way to approach this issue is to ask how the worker himself would structure such an insurance plan if he were aware of the risks and if he had to sacrifice some of his wages (on an actuarially fair basis) to obtain additional compensation. The worker ideally would provide himself with *ex post* compensation up to the level where the last dollar of benefits provides the same increase in welfare as would higher wages if he were healthy.

If injury costs were purely financial, such insurance would provide for identical income levels regardless of one's health status. However, an accident also may affect the welfare one can derive from any given expenditure level. It is usually assumed that for any particular expenditure level, a health impairment reduces the incremental increase in welfare of further expenditures.[11] In such a case, a person in ill health needs a lower overall income to provide the same incremental benefit as expenditures if he were healthy. An analogous result applies to bequests for families of workers who are killed. From the standpoint of efficient insurance, the income support under workers' compensation consequently should be partial. State plans generally provide replacement of up to two-thirds of the disabled worker's wages which, given the favorable tax status of these benefits, leads to substantial, but usually partial, earnings replacement.[12]

Unfortunately, we have no empirical basis for ascertaining the optimal reduction in income after a disability or death. Once considerations such as society's altruistic concerns are introduced, the issue becomes muddled even further. While blue-ribbon panels repeatedly call for more complete income replacement, these urgings have not met with great success, perhaps in part because of the difficulty of identifying the unambiguously ideal level of compensation.

A frequently neglected role of workers' compensation is its effect on workplace health and safety. The presence of this compensation tends to decrease the unattractive aspects of hazardous jobs, boosting the total supply of workers to risky jobs as well as the overall level of injuries in the economy. Offsetting this effect are any incentives for safety created by the funding of these programs. If a firm's insurance premiums are linked to its health and safety record, the firm will receive a financial benefit if it can improve its record. Very large firms, which account for 15 percent of all those covered and 85 percent of all workers' compensation premiums, either self-insure or are merit-rated to some extent.[13] Smaller enterprises are taxed on the basis of the performance of their industry group. Since any individual firm's performance has a negligible effect on the performance of its group, the lack of a more effective merit-rating system diminishes the possible incentives for improved safety.

Merit-rated compensation serves a twofold purpose: it provides income insurance for workers and promotes more efficient levels of health and safety. Indeed, if the market levels of risk are too high, it may be desirable to increase benefits above their efficient level or to levy an additional accident-related tax on firms to promote safety. If, however, merit rating is weak or nonexistent, the optimal level of benefits will be lower, since not only will there be no incentives for the firm to enhance safety, but there will also be the problem of workers finding hazardous employment more attractive because of the subsidy given to unsafe firms. If not merit-rated, workers' compensation may make workers worse off by subsidizing unsafe firms and consequently increasing their level of operations, raising the overall risk in the economy.

Proposals for society to subsidize workers in hazardous jobs to compensate them for the adverse effects of their jobs are similarly flawed. Although well-meaning, such a policy would have the unintended effect of increasing the number of workers employed in hazardous jobs.

A more productive direction for workers' compensation policy is to bolster the merit-rating system to enhance the incentives for safety. Proposals along these lines are usually dismissed on the grounds that small firms might be unduly burdened. If, for example, a small enterprise has an aberrationally high level of accidents, its premiums might escalate, even though the bad record is not indicative of the firm's underlying safety level. For larger firms, this problem is diminished since the larger sample size increases the accuracy of the accident record as an index of workplace conditions.

Although small firms obviously pose special informational problems of this type, more complete merit rating is still feasible. All firms of comparable size in an industry with similar accident records could be placed in one risk category. Such a system would not be tantamount to self-insurance, since firms would be assessed according to the average long-run risk of its particular group. For example, on average a small firm with one accident in its first ten person years will have a much lower long-run accident record, once the influence of the aberrationally high initial level is offset. Firms could be charged on the basis of this long-run risk, and these group-based incentives could vary with the extent of the company's premium in much the same way as deductible and coinsurance provisions limit out-of-pocket expenses of those insured. In effect, workers' compensation premiums should be adjusted to insure small firms against the risk of a large payout for an aberrationally high accident level.

Although such policies should be given greater attention, they do have important limitations. First, and perhaps most important, is that workers' compensation provides little coverage for those ailments whose cause is difficult to monitor and which also pose the greatest difficulties for worker choice. If the cause of an illness cannot be linked to a particular job, the worker may not receive any benefits. For chronic ailments that take decades to become apparent and that may have no readily identifiable cause, this shortcoming is especially acute. The firm's liability under workers' compensation will not generate incentives for improved health and safety if its record is not an accurate reflection of its performance. Since accidents are monitored more precisely than work-related illnesses, workers' compensation policies are better suited to providing incentives for safety than to promoting efficient levels of health risks.

A second and more fundamental problem is that mandated

workers' compensation levels eliminate these insurance benefits as a market-determined component of the pay package, except if a firm chooses to augment the benefits. Elimination of these benefits may be of considerable consequence, since the level of benefits an employer provides voluntarily gives the worker another form of information in much the same way as guarantees indicate higher-quality consumer products. Since the costs of such compensation are less in a safer enterprise, a worker might infer that a firm was relatively safe if it provided extensive coverage for workers injured on the job. This form of information is suppressed when coverage levels are predetermined. The extent of the loss has not been ascertained, largely because the existence of a mandatory program in all states provides no empirical basis for assessing the performance of a voluntary system.[14]

Much of the remainder of this book is concerned with government efforts to regulate risks directly through setting standards for health and safety. Although this form of regulation has dominated policies of OSHA and other agencies, it is in no way superior to the information provision and workers' compensation policies discussed in this chapter. Since different policies address different aspects of the inadequacies of the market, the optimal policy mix will undoubtedly call for several types of government action.

6

THE VALUE OF LIFE AND LIMB

TO ENSURE EFFICIENT ALLOCATION OF SOCIETY'S RESOURCES, WE should pursue only those risk regulation policies that are as beneficial as alternative uses of these resources. The unavoidable comparison is how much we should value improved physical well-being as opposed to energy conservation, greater economic growth, or some other outcome. Rather than attempt to analyze every possible tradeoff by establishing the relative value of risk reduction in terms of each other matter of concern, a simpler approach is to put these tradeoffs in terms of a common metric. Dollar values for economic outcomes serve as the usual basis of comparison and, in the case of risk regulations, they reflect society's total willingness to pay for the risk reduction.

My focus in this chapter is not on appropriate levels of compensation for the injured or for the families of workers who are killed. These values are often set at quite low levels, in part because life and health involve irreplaceable and nontransferable commodities. The more pertinent policy issue is how we should value the prevention of these adverse outcomes, which is what I will consider here.

I maintain that efforts to obtain an elusive value-of-life number are largely misdirected. Instead, analysts should be concerned with differences in individuals' values of reducing the risk of adverse health outcomes and the implications for government policy of the resulting distribution of risk-money tradeoffs. I will not become involved in the more subtle value debates as to who should be making such lifesaving decisions or whether life-saving should be targeted at the lives of the poor.

Even if we exclude these important, but less fundamental concerns, many ethical issues remain. Despite the growing economics literature on the value of life, most policymakers consider it intrinsically immoral to design and administer risk regulation policies on the basis of explicit tradeoffs between costs and health. Sanctimonious claims are often made that lives must not be bartered for dollars. Although this may be effective political rhetoric, these claims do not accurately reflect political reality. Tradeoffs are inevitable so long as resources are limited and policy initiatives that affect health favorably remain available. By making these decisions in an explicit and efficient manner, society can save more lives and prevent more illnesses and injuries than if policy implications are sketched in only the vaguest terms to disguise the tradeoffs involved.

CONCEPTUALIZING VALUE-OF-LIFE ISSUES

Although risk regulation policies will prevent some expected number of adverse health outcomes, these effects are usually probabilistic. A carcinogen standard might prevent an average of one case of cancer annually in a plant of 10,000 workers exposed to the risk. One could say that the probability of cancer is reduced after the fact from one to zero for the particular worker who would have been affected, but when the regulation is adopted this individual usually cannot be identified. Instead, the risk of cancer may be reduced by 1/10,000 for each of the 10,000 workers. The appropriate issue then for policy evaluation is not how much we value preventing certain adverse health effects, but what value we should place on small reductions in the probabilities of these outcomes for large numbers of affected workers.

This distinction is important for two reasons. First, society's attitude toward saving identified lives appears to be quite different from the attitude toward saving statistical lives. The value society places on an identified life, such as a trapped coal miner, is likely to be substantially greater than the implicit valuations of life and health status of individuals who cannot be identified, such as the prospective beneficiaries from improved ambulance service or flood control program. In the latter instance, the policy has a probabilistic effect on the well-being of large numbers of individuals. The lives that have been extended or improved may not be identifiable even on an *ex post* basis; the prevention of a flood, for example, provides no information on who would otherwise have died.

A second reason for focusing on the valuation of small reductions in risk is that the individual's relative valuation of risk reduction is likely to be greater for small changes in the risk. Since a person's willingness to pay per unit of risk reduction decreases with his wealth, he will be willing to spend relatively more for initial than for subsequent reductions of risk since his assets become depleted with each successive purchase of a decrease in risk. When extrapolated, this relationship implies that he will spend less per unit of risk to prevent certain death and injury than to bring about small changes in these risk levels.

I will refer below to implicit values of life or injury, but these terms are somewhat misleading since they pertain to the relative value of reducing risks by small amounts, not to the value of avoiding certain events. The appropriate concept for evaluation is how much individuals and society at large are willing to pay for these small reductions in risk.[1]

An alternative to this willingness-to-pay approach is the discounted present value of one's earnings, that is, the value of one's future earnings stream after appropriate weighting (using interest rates) to account for the lower value of income received in the distant future. Although this is frequently labeled the human capital approach, it has never been espoused by a prominent exponent of human capital theory because no meaningful conceptual basis exists for linking a person's future earnings to the value he would place on his life if he were faced with a lottery on life and death.

Attempts have been made to do this, and if we make very strong assumptions about individuals' preferences for income and health, we can derive an explicit linkage. The fundamental drawback of these highly structured approaches is that there is little justification for linking people's attitudes toward small health risks to their income. Greater wealth will, of course, generally decrease one's willingness to incur health risks. But even if we impose additional restrictions, such as the widespread preference for being healthy rather than not, we cannot make any judgment whatsoever as to whether it would be more rational for a middle-income person to spend ten cents or ten dollars to avoid a one-in-a-million chance of being killed.

As one might expect, equating the value of life with one's future income often has unacceptable implications. For example, it ignores the possibility that many people confronted with a life-or-death choice might try to earn more income, perhaps even illegitimately, to preserve their lives. Considerably more disturbing is that the nation's elderly and housewives would fare particularly poorly if lifesaving policies were based on their prospective market earnings.

Since some analysts have advocated an income-based measure of the value of life, it is not surprising that many critics of this line of work view discussions of the economic value of life as morally reprehensible.[2] This controversy would be muted considerably if it was understood that the economic value in question should be based primarily on the individual's own willingness to pay for small reductions in risk rather than on his contribution to the GNP or some other allegedly pertinent "economic" measure.

PROCEDURES FOR ESTIMATING
THE VALUE OF LIFE AND LIMB

The method used to assess individuals' willingness to pay to avoid death or injury is similar to that used for obtaining market prices of other components in benefit-cost analyses. One technique for assessing these prices is to use actual market behavior by, for example, estimating statistically the determinants of workers' earnings and in particular by estimating the impact of job hazards on earned income. This technique assesses individuals' tradeoffs between risks and money indirectly on the basis of preferences revealed through their actions.

An alternative approach is to ask individuals hypothetical questions to elicit implicit valuations of life and health.[3] Both procedures focus on the same risk-money tradeoff. However, the market-oriented approaches use statistical techniques to identify this tradeoff based on actual market behavior, while the interview studies ask individuals quite directly what their subjective tradeoff rates are. My own efforts have focused on estimates based on market behavior, since actual economic evidence provides a potentially more reliable basis for analyzing preferences than do hypothetical questions.

The greatest limitation of the interview approach is that respondents have no incentive to give thoughtful or honest answers, and the process of thinking about choices involving small probabilities is notoriously difficult. A person given information about a hypothetical lottery clearly has less of an incentive to evaluate his preferences regarding these risks than he would if he were incurring the same risks daily in his place of employment.

Moreover, even if the respondent has given the issues careful consideration, he has no reason to reveal his preferences honestly. He may give the response he believes will create a favorable impression on the interviewer, especially when the life to be saved is not his own but that of an anonymous member of the community. A person's altruism is likely to be much greater when he knows he will not have to back up his statements with out-of-pocket contributions. Finally, individuals may misrepresent their preferences if they believe their responses will affect the benefits they will receive or the taxes they must pay to support a public program to save lives; this is the familiar strategic issue. The fundamental and pervasive nature of these limitations suggests that interview results might best be used to supplement rather than supplant estimates obtained from market behavior.

EMPIRICAL EVIDENCE ON THE VALUE OF LIFE AND LIMB

Although the number of interview studies has not been large, the value-of-life literature using labor market data has grown considerably in recent years. A comprehensive survey would sacrifice depth for breadth. Instead I will examine in more detail a major segment of this line of work, the first four successful labor market analyses of this type;[4] the findings span the range of existing value-of-life estimates. These studies also include the first empirical estimates of the value of nonfatal injuries as well as the first estimates of heterogeneity in the value of life.

The interview studies undertaken thus far have focused on the value of life rather than other health outcomes. The first of these was Acton's (1973) investigation in which individuals were asked their valuations of different programs to save the life of heart attack victims. These interviews suggested a value of life in the range of $56,000 to $86,000 in 1980 prices. For several reasons, Acton's results are not comparable with those for the labor market, discussed below. First, the lives are post–heart attack lives and consequently should be valued less highly. Second, the sample size was rather small (thirty-six) so the estimates may not be very reliable.[5] Third, the inherent shortcomings of the interview technique make the responses possibly unreliable indicators of the individual's actual preferences.

Quite different value-of-life estimates are obtained from labor market evidence, as summarized in Table 6.1. Although the studies differ in a number of respects, particularly with regard to the risk variables used, they shared a common approach.[6] In particular, each one examined relatively large samples of data on individual earnings in an effort to estimate the risk-wages tradeoff implied by worker behavior.

The only studies that successfully estimated premiums for nonfatal injuries were my analyses, summarized in the final two columns of the table. For the two sets of survey data the results were quite similar, with the range of estimates implying that

TABLE 6.1. SUMMARY OF RISK PREMIUM STUDIES

Study characteristics	Thaler and Rosen (1976)	Smith (1976)	Viscusi (1979)	Viscusi (1981)
Sample	Survey of Economic Opportunity	Current Population Survey	Survey of Working Conditions	Panel Study of Income Dynamics
Sample size	907	3,427	496	3,977
Risk variables	Society of Actuaries' occupational death risk	BLS industry death risk	BLS industry fatal and nonfatal risk, self-perceived hazards	BLS industry fatal and nonfatal lost work-day risk
Mean death rate	.001	.0001	.0001	.0001
Mean nonfatal risk rate	—	—	.03	.03
Other job characteristics	Industry and occupation dummy variables	Industry dummy variables	Nonpecuniary characteristics of job; occupation dummy variables	Occupation dummy variables
Implicit value of life (1980 prices)	$500,000	$2.8 million and $6.4 million with 1967 data	$2.5–$3.3 million	Range, $4 million at risk of 6×10^{-4}
Implicit value of nonfatal injuries and illnesses (1980 prices)	—	—	$12–$21 thousand	$27–$30 thousand

workers implicitly valued injuries between $12,000 and $30,000 (1980 dollars). This similarity is striking since the injury risk measures were quite different; my study using the Survey of Working Conditions used the pre-OSHA injury rate reported on a voluntary basis, while my study with the Panel Study of Income Dynamics used the post-OSHA lost-workday injury and illness rate reported by all firms. In each study the annual risk of injury was about 1/30 per year. Since the injury value estimates for comparable equations differ by less than $10,000, we can be more confident of the general magnitude of these values than we could if the studies had yielded very dissimilar results. It is also noteworthy that for the Survey of Working Conditions data, which included information on workers' perceptions of risk, the annual risk premium was estimated to be $900, using both an objective risk measure and the worker's own risk assessment.

To see how wage premiums for risk are translated into implicit values for injuries, consider the following example. Suppose that a worker receives $500 annually in return for incurring a 1/30 chance of being injured. The implicit value of preventing an injury can be viewed in two ways. First, we can simply divide the risk premium by the level of risk to obtain the dollar value per unit of risk of job injury. Dividing $500 by 1/30 yields an implicit value of $15,000 for an injury. The second approach is to consider a group of thirty workers, each of whom receives $500 for facing a 1/30 injury risk, so on average one worker will be injured. The total risk premiums for these workers is $15,000 (30 workers × $500 premium per worker), which is the implicit value to the entire group of preventing one expected injury.

Notwithstanding the health value terminology, these figures simply serve as an index of the rate per unit of risk at which workers are compensated; they do not imply that a worker would accept certain injury for $15,000. In general, he would require more than this amount. Suppose the worker accepted an initial small risk of injury in return for a $500 premium. This additional income would boost his wealth level, thus raising the value of the premium he would require for the next incremental increase in risk. The opposite result holds for reductions in injury risks. Although a worker might be willing to pay $500 to reduce his risk by a small amount, each purchase of an incremental reduction in risk will deplete his wealth, thus reducing the amount he can pay for the next incremental risk reduction. Prevention of certain injury consequently will be worth less than $15,000. These interpretive caveats for implicit values of injuries are equally applicable to the value-of-life estimates below.

The implicit values that have received the greatest attention are those for life itself or, more specifically, for extending lives by preventing workplace fatalities. The value-of-life estimates range from $500,000 to $4 million, as summarized in Table 6.1. These differences do not stem from a weakness of any of these particular studies but from differences in the risk variables used and in the mix of workers included in the analysis.

Only one of these studies, Thaler and Rosen (1976), utilized an occupational risk measure that was then matched to the occupation of the workers in the sample. They used the Society of Actuaries' incremental death risk for thirty-seven high-risk occupations. This variable reflects the death risks of the occupation per se as well as death risks unrelated to work but correlated with the characteristics, life styles, and income levels of people in different occupations. The resulting patterns of risk are surprising. Cooks face three times the death risk of firemen, elevator operators face twice the death risk of truck drivers or electricians, waiters face sixty-seven times the death risk of linemen or servicemen, and actors face a higher death risk than fishermen, foresters, power plant operatives, and individuals in many other more physically demanding occupations. The overall risk of death for the sample averaged 1/1,000.

In contrast, the studies by Smith (1976) and myself used the BLS average death risk for the worker's industry. The major weakness of this measure is that not all jobs in an industry pose the same risk. For my work with the Survey of Working Conditions, I also took into account whether the worker perceived some risk in his particular job.[7] Although combining information on workers' perceptions with industry risk levels may improve the accuracy of this variable, none of these measures is a fully adequate substitute for information on the worker's perceptions of the probabilities of different health outcomes.[8] In the absence of this information, it is not possible to ascertain whether risk measures based on the industry risk or those based on the occupational risk are more accurate.

Since the value-of-life estimate with the occupational risk measure is only $500,000 (in 1980 prices), as compared with $2 million or more for the various industry risk measures, the natural question is, which is correct? Some analysts have dismissed value-of-life estimates in the millions as unreasonable. Others, including many strong advocates of risk regulation, find the high estimates as the least offensive results in what they view as an unethical line of research. On the basis of economic theory, we cannot say whether the correct value of life should be about $500,000, $1 million, or $6 million. Certainly the views of outside analysts or government officials as to what these premiums ought to be should not be pertinent. What matters are the risk preferences of the workers themselves, and their preferences may be consistent with widely varying values of life.

Claims that the high value-of-life estimates are unreasonable are usually supplemented with an alleged example of their implausibility, such as the following. Suppose that a worker with an income of $20,000 and an implicit value of life of $1 million is told that he has an illness that poses a 1/5 chance of death. One might then calculate that he should be willing to pay $200,000—or ten times his annual income—to eliminate this risk. The implications of this example appear unreasonable because risk-money tradeoffs for small changes in risk are being extrapolated to large risk reductions. Since the implicit value of life declines as one's resources are depleted, failure to reduce the applicable value-of-life number when evaluating large risk reductions will always lead to such overstatements of the individual's willingness to pay for such decreases in risk.

HETEROGENEITY IN THE VALUE OF LIFE

The different value-of-life estimates are not necessarily incompatible, because individuals may differ substantially in their willingness to incur risks. These differences will be reflected in the jobs they select, and differences in the mix of workers considered might result in quite different risk-money tradeoffs.

In most market situations, these types of issues never arise, because market prices are set identically for all transactions. However, unlike standard consumer items, death risks do not command a single price. The risk is inextricably linked to the job; it cannot be divided to yield a constant price per unit of risk. Those individuals who are least averse to such risks are willing to accept a lower compensation per unit of risk than the rest of the working population. As a result, they are inclined to accept larger risks with lower wage premiums per unit of risk. Workers in the high-risk sample used by Thaler and Rosen in effect have shown themselves to be less averse to death risks than the rest of the population. The role of this self-selection process is not as significant for the other studies listed because these samples were not generated on the basis of the riskiness of the positions. The workers in the Thaler and Rosen sample faced an average annual death risk of 1/1,000, or about ten times that of the workers in the other samples, who faced a risk comparable to that of the typical worker.

Since different people attach different values to their lives, empirical analyses should not be directed at estimating an elusive value-of-life number. Rather they should estimate the schedule of values for the entire population. In Figure 6.1, the line VL illustrates such a schedule. As the percentage of the population incurring the incremental risk increases, so does the marginal valuation of life. Those who price their life the cheapest are drawn into the market first; higher wages must be paid to lure additional workers into risky jobs. Thaler and Rosen focused on the lower tail of the population, those who appeared to value their lives at approximately $500,000. More representative samples yielded higher values such as $3 million. Empirical estimates can most accurately be viewed as weighted averages of points along the marginal value-of-life curve for the population. If the sample contains a disproportionate number of workers from the occupations posing substantial risks, it generates lower average values of life and limb than if the sample is more representative. Ideally, we would like to ascertain how risk premiums

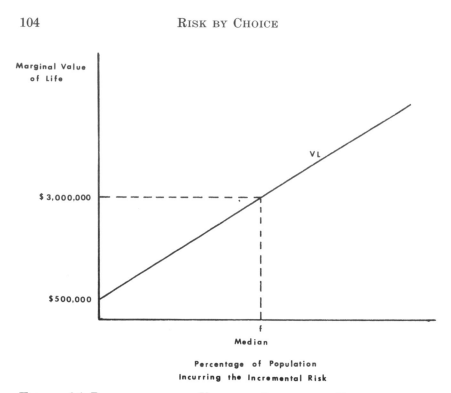

FIGURE 6.1. RELATION OF THE VALUE OF LIFE TO THE PERCENTAGE OF
THE POPULATION INCURRING THE RISK.

per unit of risk vary with the level of risk. This information
would enable us to ascertain the shape of curves such as VL.

My most recent study, which is summarized in the final column of Table 6.1, focused on the heterogeneity of the value of
life. The sample of workers analyzed consisted of almost 4,000
full-time workers, who faced an annual risk of death of .0001 and
a risk of nonfatal injury of .03.[9] This large sample made it possible to estimate the variation in the risk premiums for fatal
risks.[10]

While there was no evidence of any variance in premiums for
injuries and illnesses, there were significant differences in the
value of life for workers at different risk levels, as one would expect based on the three studies. The findings in Table 6.2 indicate
the variation in these values for two different equations.[11] The
different quartile risk levels shown pertain to the entire private

TABLE 6.2. HETEROGENEITY IN THE VALUE OF LIFE

Risk level	Annual death risk[a]	Value of life	
		Wage equation	Log wage equation
First quartile	0.48×10^{-4}	10.1×10^6	6.3×10^6
Second quartile	0.54×10^{-4}	10.1×10^6	6.3×10^6
Third quartile	0.70×10^{-4}	$ 9.8 \times 10^6$	6.2×10^6
Fourth quartile	6.18×10^{-4}	$ 3.7 \times 10^6$	3.5×10^6
Sample mean	1.04×10^{-4}	$ 9.5 \times 10^6$	6.0×10^6

Source: Viscusi (1981).
a. Risk levels for each quartile pertain to the upper boundary of the quartile.

nonfarm economy rather than to the particular sample analyzed. Thus the safest quartile of workers faces an average death risk of 0.48×10^{-4} or more. Those individuals at the average risk level for this range have an implicit value of life of $6 million and $10 million for the two equations. It is especially striking that the value of life estimates for the first and third quartile differ by only 2–3 percent, which indicates that three-fourths of all workers appear to have very similar values of life. There is a dramatic decline in the value of life as one considers workers in the most hazardous quartile. The risk for these workers is 6.18×10^{-4}, which is associated with a value of life of just under $4 million.

Although even this estimate is above that obtained by Thaler and Rosen, the risk levels are lower as well. Since the value of life falls off very steeply as the risk level is increased, low estimates such as theirs are not implausible. The first equation yields a negative value of life at the risk level for their sample, while the second yields an estimate of $800,000. Although estimates such as these are not likely to be reliable since they are outside of the risk range for most of the workers in the sample, they do suggest that all of the disparate value-of-life estimates may not be contradictory at all if one recognizes the heterogeneity of the values being analyzed.

These results suggest that there are important differences in the value of life for different people. One should, however, be cognizant of the limitations of this and other studies. Since industry risk measures were matched to workers based on rather broad classifications of the worker's industry, the estimates for the Panel Study of Income Dynamics may not be as reliable as those in my earlier analysis, in which it was possible to match the risk measure to very refined industry classifications and to take into account workers' risk perceptions and the impact on wages of other job attributes. Similar problems of comparability make it infeasible to rank the relative accuracy of the findings by Smith and those by Thaler and Rosen.

However, one can safely draw two principal conclusions. First, the heterogeneity in individuals' willingness to pay for risk reduction appears to be quite substantial. If they incur these hazards voluntarily, those facing the greatest risk will largely tend to be individuals with relatively low values of life. Second, the general order of magnitude of workers' value of health effects is fairly large. The value of preventing nonfatal job injuries is about $20,000 to $30,000, while the value of life ranges from about $500,000 for workers in high risk jobs to $3 million or more for more representative workers. For the purposes of policy evaluation, the Thaler and Rosen estimates of $500,000 seem most reasonable for workers in high-risk jobs (about 1/1,000 annually). For workers facing less severe risks (about 1/10,000 annually), an estimate of $2 million appears more reasonable. This figure is based on my analysis of data from the Survey of Working Conditions. This survey included the most extensive job characteristic data and the most refined information on the worker's industry, facilitating the process of creating an accurate measure of the job risk using data on industry-wide risks. Although these estimates clearly can be refined through further research, they do provide a more reliable basis for formulating risk regulation policies than do unfounded speculations by policymakers regarding the appropriate value of health benefits.

POLICY APPLICATIONS

The most immediate significance of the empirical results is their implication for labor market performance. If workers were not compensated adequately for the risks they incurred, one would conclude that the market did not function effectively, perhaps because of systematic individual misallocations. The theme of inadequate compensation runs throughout the more sociologically oriented literature on occupational safety.

As my empirical results indicate, the annual compensation for all job risks averages $900 per worker. Unlike stuntmen and other workers who received clearly significant hazard premiums, a typical worker in a hazardous occupation does not receive enough additional remuneration to be obvious to the casual observer. It is also important to note, however, that the risks incurred are not very large; the probability of a fatal injury is only about 10^{-4}. To ascertain whether the amounts accepted by workers for additional risks are small enough to suggest some form of market failure, one should examine not the absolute level of risk compensation, but the implicit values of life and of injury. The empirical results indicate that the magnitudes of these values are quite impressive—in the millions for fatalities and on the order of at least $20,000 for injuries. Although there is no way to ascertain whether these values are above or below those that would prevail if workers were perfectly informed, the magnitudes are at least suggestive in that they indicate substantial wage compensation for job hazards.[12]

These findings do not imply that the government should not intervene. They do indicate, however, that it is doubtful that one can base the case for intervention on the absence of compensation for risks of death and injury. The estimated values of life and limb also can be used in assigning dollar values to the impacts of occupational health and safety regulation. If safety standards reduce the perceived risks faced by workers, their wages will fall in a competitive market.[13]

More generally, the empirical results set forth here have important implications for policy analyses of projects involving health risks. To date, there has been little systematic attempt to incorporate dollar valuations of life and limb into these analyses. Even federal evaluations of water resources projects, which assign somewhat arbitrary values to recreation benefits and other project impacts, ignore the lifesaving consequences of flood control and include only reduced property damage in the tally of benefits and costs. Much of the problem derives from society's reluctance to make explicit the tradeoffs between dollars and lives. People are likely to say, "if additional expenditures can save lives, we will spare no expense in doing so." Although this maxim is not entirely implausible when dealing with identified lives, it clearly does not reflect the reality of public decisions or common medical practice or, for that matter, of private decisions. Public decisions concerned with individual welfare implicity assign a finite value to life and other health outcomes except in the rare instances in which additional expenditures would accomplish nothing. Ignoring the issue of valuation of life and limb may circumvent the problem of offending people's sensitivities when one makes the tradeoffs explicit. But it may be very costly in that it sacrifices lives that could have been improved or saved by a more systematic allocation process. An important issue for society as a whole, and one that many people are unwilling to face, is whether lives should be sacrificed in the effort to maintain the illusion that we will not trade off lives for dollars.

A possible alternative to explicit valuations of life and limb is the use of cost-effectiveness analysis. Instead of ascertaining which program offers the greatest net benefits to society, one estimates the costs per life saved through different programs and allocates funds where they are most productive. Although this approach is useful in highlighting clear-cut cases of inefficiency, it has important limitations. First, suppose that OSHA spends $2 million for each life saved through its chemical labeling regulations, the Medicare program spends $4 million per life saved, and nuclear power plant safety regulation costs $16 million per life saved. Additional lives could be saved if funds were reallocated to make the cost per life saved the same across different programs. Even if such a reallocation were made, however, the policy might not be optimal, for there are no means to determine the optimal level of expenditures in the life-extending area.

The second shortcoming of cost-effectiveness analysis is that it provides no guide to action when there are a variety of project impacts of unknown value. Programs that extend lives typically have other health effects, such as influencing the probability of illness and the well-being of those who are ill. In such instances, one cannot summarize a program's effect by saying that it costs X dollars per life saved, since it influences many health-related concerns.

Finally, analyses of costs per life saved are not meaningful if the lives saved have different lengths and different qualities. Extending the life of an elderly individual or someone in a permanent coma differs greatly in value from reducing the incidence of fatalities among healthy individuals whose lives will be greatly extended by policy intervention. In short, the extent of life lengthening and the quality of the life that is lengthened are important matters that are not readily subsumed into the simple cost-effectiveness calculation.[14]

Although policymakers can choose among programs with several qualitatively described impacts, in doing so they implicitly assign dollar values, or shadow prices, to the different outcomes. Making policy decisions on this basis raises two key problems. First, there is no guarantee that the policymakers' attitudes about the worth of life and limb coincide with those of society as a whole. The preferences of project beneficiaries, not of legislators and bureaucrats, should be of paramount concern and should not be ignored in the decision-making process.[15] Second, if quantitative values are not assigned to different policy impacts, the most productive allocations may not even be included in the list of policy options being considered. Typically, the processes of program design and decision are separated, because different groups of individuals are responsible for drawing up the menu of policy options and choosing among these alternatives. Including explicit values of life and limb in the early stages of policy design assists in ensuring that society's valuations are incorporated in the entire policy choice process.

If dollar values are to be assigned to different impacts on life

and health, the controversy centers on what these values should
be. Implicit values obtained by observing market behavior are
instructive in establishing the value of life to the individual. Society at large, however, also has a stake in the health of its members. The group most affected by the external effects of death or
illness is the individual's family. To the extent that a worker
takes into account the preferences of other members of the household when making his employment decision, the market estimates reflect such externalities. Although the outcome might not
be exactly what one would observe if the external effects were
evaluated, a substantial input of this type no doubt affects employment decisions.

The altruistic concerns of individuals other than the worker's
family are not reflected in market estimates. Unless such concerns are especially strong, however, they will not substantially
affect the value of life obtained using workers' own preferences.
Since market behavior is not instructive about such altruistic
concerns, interview studies might be used to obtain further information on this issue.

The most fundamental determinants of the appropriate value
of life are the preferences of those whose health is being endan-
gered. In analyzing the value of life and limb, one should use the
information provided by the nature of the risk incurred. In particular, voluntary and involuntary risks should be treated quite
differently. The only distinction that has traditionally been
drawn between these two types of risk is that in a market system
prices reflect individuals' value of voluntary risks to life and
limb.[16]

In the discussion below I assume that the pertinent schedule of
values for the population has already been obtained through previous empirical work. The key question is how to use these schedules in policy evaluation in situations in which individuals incur
risks voluntarily and involuntarily. Although the degree of voli-
tion spans a continuum of possibilities, for simplicity I focus on
the polar cases of completely voluntary and completely involuntary risk.

The nature of the risk conveys important information about

the implicit value of life and limb being assigned by the affected population. Other things equal, it is those who place the lowest dollar value on the expected loss to their health who choose to incur the risk. If individuals choose to live in a flood-prone area, to drive cars while intoxicated, or to work at hazardous jobs, the government's assessment of the value of the health gains from safety regulation should be quite different from its assignment when no element of free choice is involved.

Suppose, for example, that the characteristics of the affected population are comparable to those of the sample of workers used to estimate the heterogeneity of the value of life. The median value of life corresponds to that of the individual at point f in Figure 6.2. Consider the situation in which the risk is involuntary and affects the whole population. Then the use of the median individual's value of life gives correct estimates of the benefits of lifesaving activities if the value of life schedule is linear,

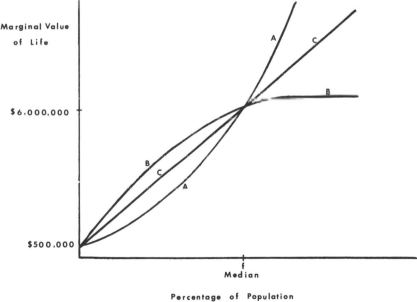

FIGURE 6.2. POSSIBLE SHAPES OF VALUE-OF-LIFE SCHEDULES.

shown by the line CC; overstates the benefits if the schedule is concave, like BB; or understates the benefits for schedule AA. If the risk is voluntary and affects only half of the population, the use of the median individual's preferences at point f overstates the value of life irrespective of the shape of the value of life curve, since all other individuals who have chosen to incur the risk value their lives at less than the amount for the median individual.

Curve BB in Figure 6.2 has been drawn using the log wage equation from the earlier empirical results on the heterogeneity of the value of life. For these estimates the marginal value-of-life curve rises steeply at first, then becomes relatively flat. If these results accurately reflect the distribution of the value of life, then using the value of life for the individual at the median risk level will overstate the average value of life, whether the entire population is affected or simply those who have chosen to incur the risk.

For purposes of policy evaluation, it is the preferences of the average individual that are of consequence. The value of health-enhancing program benefits is computed by multiplying the number of lives saved or injuries prevented by the average value attached to these outcomes. In contrast, market outcomes reflect the preferences of the marginal worker. The valuation of life and limb of the worker who accepts the risky job and who is most averse to the hazards is instrumental in setting the wage rate, not the preferences of all the other inframarginal workers who would be willing to accept less than the going wage for the hazardous job.

Much of the bias against the value-of-life approach is based on a misunderstanding of the likely consequences of formulating policies in this manner. Contrary to widespread belief, policies based on appropriate values of life do not necessarily provide for less lifesaving activity than those judged on the basis of risk levels alone. In a recent case involving airline safety, the Federal Aviation Administration chose not to require wing modifications in the accident-prone DC-10 because doing so would diminish the accident risk by only one chance in a billion. Ignoring low-probability events such as this is not always desirable on a benefit-cost basis, however. To justify the repair cost of approximately $2,000 per plane, one must show that the expected value of the risk reduction exceeds this amount. If the plane makes 100,000 passenger trips per year (say, 1,000 flights with 100 people), then over the life of the plane the expected value of risk reduction exceeds $2,000, provided that the value per life saved is at least $2 million, which is not unreasonable. Policymakers may dismiss relatively small risks of catastrophic consequences in situations in which the benefits of the risk reduction are quite substantial. Ignoring such valuation issues altogether does not necessarily promote greater risk reduction than does the more balanced approach.

Although much research remains to be done in these areas, policymakers should not be reluctant to use the dollar estimates available, which are no less precise than most other dollar values assigned to project impacts in benefit-cost analyses. Moreover, the policy decision may not be very sensitive to whether the correct average value of life for the affected population is $1 million or $3 million, whereas substantial welfare losses may result if the value assigned to life is only $100,000 or if the problem is ignored altogether. Even when used in an appropriate manner, dollar values for life and limb may offend some individuals' moral sensitivities. A greater danger is that these tradeoffs will be made without systematic analysis. Society may then be paying a substantial implicit price in lives sacrificed in an effort to preserve popular illusions.

7

HOW TO SET STANDARDS
IF YOU MUST

IF SETTING STANDARDS WERE TANTAMOUNT TO DICTATING levels of health and safety, one could apply the guidelines from the past two chapters to assess the costs and benefits of different risk levels, then select standards policies accordingly. In practice, the level of the standard and the actual risk outcome are quite different, since the enforcement of standards is seldom completely effective. Particularly for OSHA standards, which are enforced so weakly that noncompliance is rampant, the context of standard setting exerts a profound influence on the implications of different policies.[1]

Enterprises will choose to comply with regulatory standards if the expected costs of noncompliance exceed the expected costs of compliance. If noncompliance is less costly, a firm will risk the penalty, then make the mandated changes in workplace conditions only after being inspected and threatened with an ever-increasing schedule of fines for noncompliance.

In an ideal world of full compliance, stringent standards, such as those linked to the lowest technically feasible risk level, yield the greatest safety gains, but in practice they may have quite different effects. For firms that choose to comply, tighter standard levels enhance safety; a counterbalancing influence is that more stringent standards raise the relative costs of compliance, leading more firms to ignore the program and to leave their workplace conditions unchanged. The net effect depends on the relative strength of these influences. In the extreme case of standards that are subjected to widespread public ridicule, such as OSHA's portable toilet standard for cowboys, the risk to a firm of adverse publicity from noncompliance is reduced, and the incentive to invest in workplace health and safety is diminished even further. Even if one advocated that OSHA attempt to promote but a single objective, worker health and safety, it would not typically be desirable to establish standards at the most stringent, technically feasible levels.

An analysis of the likelihood of compliance should be incorporated in any assessment of the impacts of different policies. Other factors besides the level of the standard also influence compliance; firms are more likely to comply as the cost of meeting the standard decreases and as the expected penalties for noncompliance increase. Enforcement policies such as OSHA's, which have a very weak enforcement mechanism (low penalty levels, infrequent inspections, and few violations per inspection), consequently do very little to enhance worker safety.

With any uniform standard, the firms that are affected are likely to show a diversity of responses; there will be compliance where the costs of meeting the standard are low and noncompliance where these costs are high. Recognition of this heterogeneity may lead to greater compliance and worker safety. In particular, the standard should be more stringent where the compliance costs are very low. Policymakers might, for example, set different standards for different industries, for different stages of production, or for facilities of different ages. Heterogeneity in the level of benefits from a standard, as in the case of groups that are particularly sensitive to some occupational exposure, perhaps because of synergistic effects with other risks, similarly should lead to differentiation in the level of the standard.

The next step in evaluating policies is to assess the benefits and costs of different standards levels by converting the health benefits into dollar equivalents, as outlined in the previous chapter. There are two interrelated guidelines for an optimal standard; it should provide the greatest net gains to society; that is, it should maximize benefits less costs. To achieve such an optimal policy in the usual situation in which the incremental (or mar-

ginal) benefits from tightening the standard are decreasing and
the marginal costs of compliance with tighter standards are increasing, one should tighten the standard until the additional
benefits no longer exceed the additional costs.

These basic principles ideally should also reflect the heterogeneity of costs and benefits associated with the standard. If compliance costs are less for new facilities than for existing facilities
that must be modified, the regulatory agency should set differential standards based on these cost differences, so that marginal
benefits and costs will be equated for each type of facility. The
most prominent example of an OSHA standard that reflects this
general principle is the inclusion on the tractor rollover protection standard of a "grandfather" clause exempting existing trac-
tors. While some analysts have bemoaned the "new source bias"
in which new facilities face tighter standards, some discrimina-
tion along these lines should be encouraged as a means to make
standards policies more effective. The extent of this discrimina-
tion should reflect the relative costs and benefits of compliance,
leading to differentiation in the standards rather than complete
exemption of existing facilities. One can, for example, justifiably
question the extent to which EPA regulations have placed a disproportionate burden on new pollution sources.

Full recognition of the heterogeneity in the costs of compliance would make the standards program tantamount to a penalty system, which is generally favored by economists. Suppose
OSHA fined firms according to their health and safety performance or workplace conditions, with the level of the fine reflecting
the marginal benefits of additional improvements. Enterprises
with high risk levels or with very severe types of injuries would
pay a higher fine. Firms could decide whether or not to comply
based on the costs involved. The advantage of this approach is
that the marginal benefits and marginal costs of compliance
could be equated on a decentralized basis, ensuring efficient
levels of health and safety. A hazard penalty system of this type
parallels proposals that water pollution be controlled by imposing effluent charges on pollutors.

Under a standards program, many firms will ignore the standard altogether until identified by the regulatory agency as non-compliers. Then they face substantial penalties that are levied until they comply. Unless the standard has been set at a level reflecting the firm's compliance costs, there will be potentially large inefficiencies, as the costs imposed may far exceed any health benefits. So long as standards do not reflect the heterogeneity in benefits and compliance costs, they will inevitably create inefficiencies and unnecessary distortions.

The OSHA Carcinogen Policy

In one particularly important case, we can make a direct comparison of the guidelines discussed above and OSHA's principles for policy design. Whereas most rulemakings involve regulation of specific substances, the OSHA carcinogen policy, which was proposed in 1978 and issued in 1980, established a general set of principles for the identification, classification, and regulation of potential occupational carcinogens.[2] The open-ended nature of these guidelines is reflected in the wide range of cost estimates: $83 billion if OSHA sets loose standards for 38 substances, $296 billion for moderate standards for 1,970 substances, $526 billion for moderate standards for 2,415 substances, and prohibitive cost levels for any stringent standards, regardless of which number of substances is regulated.[3] The potentially enormous costs associated with the carcinogen policy and the importance of a well-designed policy to the well-being of workers gives this set of guidelines a preeminent role.

Unfortunately, the highly simplistic OSHA carcinogen policy bears little, if any, relation to meaningful guidelines for policy design because it abstracts from almost all of the variations in the costs and benefits of regulating different substances. Rather than promulgate rules for standards design along the lines of the principles discussed in the preceding section, OSHA established two classes of risks, Category I and Category II carcinogens, and set uniform guidelines for every substance within each category.

A substance is placed in Category I if it is a potential occupational carcinogen in 1) humans, or 2) in a single mammalian species in a long-term bioassay where the evidence is either in concordance with other scientific evidence of potential carcino- genicity or if the Secretary of Labor chooses to wave the concor- dance requirement. If the evidence of carcinogenicity is only "suggestive" or if it is based on mammalian bioassays that are not in "concordance" with other evidence, the substance is placed in Category II. Using engineering and work practice controls, firms must reduce Category I substances to the lowest feasible level or the level at which it has been shown that there is no risk. Moreover, if a suitable substitute exists, no occupational expo- sure to the potential carcinogen is permitted. Standards for Cat- egory II carcinogens are not as tight or as well defined; they will be set at "appropriate" levels on a case-by-case basis. Although no substances have yet been categorized under this policy, in 1980 OSHA identified 107 substances for possible classification.[4]

What is most striking about the carcinogen policy is that it is not designed to promote the interests of the workers or of so- ciety at large. The driving force behind it is the strength of evi- dence pertaining to possible carcinogenicity; the level of risk and the costs and benefits of regulation are not matters of concern to OSHA. Following these guidelines, one might ban substances for which there was strong evidence of a minor hazard and not regu- late substances for which there was weak evidence of a substan- tial risk. These relative priorities are certainly misplaced. OSHA should utilize whatever information is available and focus on regulating substances based on the greatest potential net bene- fits to society. Even if policymakers choose to ignore the costs of regulation, as OSHA does, they must consider the magnitude of the risk in order to determine the health benefits to the workers.

OSHA's decision to reduce Category I carcinogens to no-risk or lowest feasible levels completely ignores the types of benefit- cost tradeoffs that are critical to a balanced policy. Rather than have critical cutoffs for carcinogens that must be effectively banned, it is more desirable to assess the benefits and costs asso- ciated with different exposure levels and then set the standard offering the greatest net benefits.

The only area in which cost considerations enter explicitly is with regard to banning substances for which no suitable substitutes exist. The difficulty is that the level of costs per se should not dictate whether a substance is banned. It may be inexpensive to replace substances that pose trivial hazards, and it may be very expensive to replace very hazardous substances. In each case, consideration of the level of costs and evidence of possible carcinogenicity are not a sufficient basis for making policy. Carcinogen regulations should be based on their overall merits, not on crude rules of thumb.

Once the economic framework for assessing carcinogen policies is structured appropriately, policymakers can develop the kind of information needed to make meaningful decisions. At present, the assessment of benefits consists of little more than ascertaining the degree to which a substance can be classified as a carcinogen. With a benefit-cost approach, three types of information are required: 1) the nature of the dose-response relationship, 2) the size of the population affected by the risk and any heterogeneity in the nature of the response, and 3) an assessment of the value of the health effects—their severity, reversibility, and, if possible, their comparable monetary value.

Information about the dose-response relationship is usually restricted to the establishment of some exposure level at which the risk is not zero. Even limited information can be utilized in a more effective manner than at present by making some reasonable assumptions about the dose-response relationship. A common assumption in the medical literature on risk thresholds is that the risk is zero below the threshold and increases linearly thereafter. OSHA's obsession with eliminating all risks that pass some threshold can only be justified if the risk rises discontinuously, perhaps from zero to one, once the threshold has been reached.

Even a properly formulated threshold model may simply be a reflection of our general ignorance of the properties of different substances. When we have a small number of studies of a substance's carcinogenicity, there is a tendency to identify as the threshold the lowest level at which significant cancer risks have been observed. The underlying process generating the risk may be better characterized as a continuous one, which may have superior statistical properties.[5]

The next two types of information, the population affected and the distribution of the health effects, present fewer practical difficulties. Perhaps most important and least considered is the assessment of the severity of the health effects. If they are relatively minor or potentially reversible, the benefits from risk reduction will be much less than if workers are being killed or permanently disabled. While the evidence on the value of life and worker injuries is often instructive, there is a continuing need for refinement of the appropriate values for the diverse health effects that are the focus of risk regulation policies.

SETTING THE LEVEL OF THE STANDARD

If labor markets worked effectively, market-determined risk levels would be at optimal levels. In instances where there are inadequacies in the market, risk standards must be set by a central agency after making explicit calculations of benefits and costs. Such calculations are also instructive in the case of risks for which it is believed that no regulation is needed, since they provide a check on the underlying assumption that the market is effective.

SAFETY STANDARDS: COMMERCIAL DIVING

By issuing almost all of its safety standards without any supporting analysis, OSHA sacrificed an opportunity to ascertain the desirability of the intervention. In the case of the commercial diving standard, which was proposed in 1977 and selected for sunset review by the Reagan administration, supporting information was provided as part of the regulatory oversight process.[6] The physical hazards of diving are well known; they include drowning, gas embolism, oxygen deficiency, and related problems. The overall mortality rate of divers is quite high—about

1/250 per year. If we assume that workers in this high-risk group receive wage premiums reflecting an implicit value of life of $500,000, which is comparable to levels for workers in other high-risk jobs, the annual death risk premiums should be about $2,000. If there are also compensating differentials for other unpleasant job attributes, such as diving in cold water, the net effect should be that these workers will be paid fairly well. Divers' salaries average $29,000 to $35,000 annually, and saturation divers earn considerably more, so the overall wage levels are not out of line with what one would expect.[7]

OSHA sought to reduce these risks by issuing a safe practices manual, by requiring that a diver's log be kept, and by establishing a variety of requirements for diving practices and for the medical condition of divers. It is reasonable to assume that the regulations will save about three lives annually at an annual cost of $32.2 million, or $10.7 million, per life, an amount that exceeds the estimated value of life for even those workers who value their lives quite highly and over twenty times as large as the implicit values of life for workers in high-risk jobs.

The policy implications are twofold. First, it is not surprising that the safety provisions required by the OSHA standard were not already provided voluntarily through market forces, since it would not be efficient to do so. The costs of the provisions far exceeded the benefits to workers, as the subsequent reduction in wage risk premiums for the firm would be roughly 5 percent of the cost if workers assessed the improvement in safety accurately. The failure of the market to provide for these safety precautions consequently does not imply that there is any inadequacy in the operation of this market for risky jobs. The second and related observation is that based on any reasonable assessment of the merits of diving standards, this policy cannot be justified.

The Efficacy of Health Standards

In the case of many health risks, it is usually assumed that the market does not work effectively. But even if that is the case, one must still ascertain whether intervention is warranted. One evaluation technique discussed earlier is to analyze the cost per adverse health impact prevented and, if this amount is clearly disproportionate to the value of the health benefits or if the costs are higher than those of other programs producing these health gains, the policy should not be adopted.

Table 7.1 presents the costs per health impact associated with different proposed OSHA standards. In the case of acrylonitrile, for which OSHA proposed several options, the standard listed is the one actually adopted. I calculated the costs per case using data provided as part of the White House regulatory review process.[8] Although similar calculations are often made by the regulatory oversight group when sufficient information is available,

TABLE 7.1. COSTS OF HEALTH IMPROVEMENTS UNDER OSHA STANDARDS

Description of proposed standard	Health impact	Average cost per case
Noise: 85 decibels	Hearing loss	$169,000
Lead: 100 μg/m^3 in air	High lead levels in blood	$276,000
Cotton dust: 0.2 mg/m^3 of respirable dust in air	Byssinosis	$415,000
Acrylonitrile: 2.0 ppm in air	Cancer	$1.8 million
	Death	$4.6 million
Arsenic: .004 mg/m^3 in air	Death	$5.6 million (midpoint of range)
Coke oven: 0.3 mg/m^3 particulate in air	Death	$13.9 million

Source: Calculations by the author using data from Miller (1976, arsenic); Levine (1976, arsenic); Morrall (1976a, coke); and U.S. Council on Wage and Price Stability (1977a, cotton), (1977c, lead), (1978, acrylonitrile). The costs per case represent present values calculated in terms of 1980 prices, using a 10 percent interest rate whenever possible.

often OSHA has not provided the data needed to assess the benefits of the standard. Even more troublesome is that such calculations are not a routine part of OSHA policy design.

The figures in the final column of Table 7.1 highlight the diverse ways in which the government can allocate resources to affect individual health. With these policies, we can reduce lead in workers' blood below the levels linked with serious health effects for $270,000 per affected worker, eliminate byssinosis at $415,000 per case, reduce hearing loss at $169,000 per affected worker, eliminate cancer for $1.8 million per case, and extend lives threatened by other hazards for $4.6 million to $13.9 million per life.

Before we can allocate resources efficiently, we must make some judgments about the relative importance of these health impacts. Is it more desirable to prevent three cases of hearing impairment than to prevent one case of byssinosis? To date, OSHA has yet to make such comparisons, focusing instead on lowest feasible risk levels.

The cost figures also highlight some clear-cut inefficiencies. More lives could be saved for fewer dollars by, for example, loosening the coke-oven standard and tightening the acrylonitrile standard. (Both of these policies are relatively inefficient methods for extending lives, however.) Once we begin the process of trying to actually set the standard level optimally, we need two additional pieces of information: the value of the health impact and the influence of different levels of the standard on the benefits and costs.

A CONTINUOUS CHOICE: ARSENIC

The choice of the optimal level for the standard can be illustrated using the cost per life associated with arsenic standards of differing stringency.[9] The average cost per life saved for different standards is given in column 2 of Table 7.2, and the marginal cost per life in the third column. This column is of principal interest since the incremental cost per unit benefit should determine the stringency of any regulatory policy.

Table 7.2. Cost per life for different arsenic standards

Standard level (mg/m^3)	Average cost per life	Marginal cost per life
.10	$1.25 million	$1.25 million
.05	$2.92 million	$11.5 million
.004	$5.63 million	$68.1 million

Source: Calculations by the author using data from Miller (1976) and Levine (1976).

The marginal cost per life escalates dramatically as the standard is successively tightened, rising to a value of $68.1 million at the .004-milligram level proposed by OSHA. This cost dwarfs any reasonable estimate of the value of life. The lives being extended in this instance are those of workers in the zinc, lead, and copper smelting industries, none of which are particularly safe. In view of the workers' decision to accept jobs in these risky industries, one would expect that these workers had relatively low values of life. Even if a higher value of life—just over $1 million per worker—were used, only the least stringent alternative would be warranted.

The escalation of the marginal costs per life to levels close to $70 million and the much more modest increase in the average cost per life shown in Table 7.2 highlight the importance of focusing on incremental changes. When the costs of a tighter standard are averaged, the efficacy of the policy may look much greater than would appear if the changes in the costs and benefits were isolated.

Heterogeneous Standards: Cotton Dust and Noise

If there are wide variations among firms in compliance costs, it is often desirable to establish several standards, reflecting the differences in the relative costs per unit of benefit achieved. Although OSHA has yet to set standards optimally on an aggregative basis, much less after taking such variations into account, sufficient information is available to assess the potential gains that may be achieved.

One type of variation is within an industry. In different phases of the production process the costs of meeting the standard, as well as the benefits of doing so, may differ. The risk of byssinosis increases in the later stages of cotton processing for any given concentration of dust, because the dust is more toxic in the weaving stage than in the yarn preparation stage. The costs of compliance also vary by stage of processing.[10] The net effect of these variations is captured in the data in Table 7.3 pertaining to the marginal costs of preventing byssinosis in different stages of processing. The greatest health benefits per dollar are achieved by regulating mill slashing and weaving. For higher values placed on byssinosis, it is then optimal to regulate yarn preparation as well, after which it becomes more cost-effective to tighten the mill slashing and weaving standard. Only for very high values placed on byssinosis will it be optimal to set a very tight standard for yarn preparation.

The standard set by OSHA in 1978 did differentiate between segments of the industry, with a standard of .2 mg/m^3 for yarn preparation and .75 mg/m^3 for slashing and weaving. These standard levels do not represent a truly cost-effective solution

TABLE 7.3. THE IMPACT OF VARIATIONS IN THE COTTON DUST STANDARD

Stage of cotton processing	Marginal cost per case (thousands of dollars)		
	0.5 mg/m^3	0.2 mg/m^3	0.1 mg/m^3
Yarn preparation	56	593	6,268
Mill slashing and weaving	22	1,338	1,867

Source: Calculations by the author using data from U.S. Council on Wage and Price Stability (1977a).

since the marginal cost per case was much higher for yarn preparation. Nevertheless, there was an attempt to relax the slashing
and weaving standard to avoid the rapid acceleration in compliance costs that would have occurred with a .2 mg/m^3 standard
for that segment of cotton production.

Although the differentiation is laudable, the absolute level of
the standard is overly stringent. A marginal cost of over half a
million dollars per case year of byssinosis dwarfs any reasonable
estimate of society's willingness to pay to prevent these ailments.
Although the composition of the byssinosis-related cases was not
analyzed in OSHA's regulatory analysis, reversible grades of the
disease involving occasional chest tightness on Monday morning
comprise most cases of byssinosis prevented. Very few cases involve irreversible impacts or a substantial loss in lung function.
If byssinosis cases were valued at $22,000, a standard of
.5 mg/m^3 would be desirable for mill slashing and weaving but not
for yarn preparation. Since even this estimate undoubtedly
overstates the annual loss associated with occasional chest tightness, none of the standard levels shown in Table 7.3 appear justified on a benefit-cost basis.

Although no engineering controls appear desirable, an alternative approach using protective equipment is promising. Light,
disposable dust masks could be worn by most cotton dust workers
for only a few hours a day and would provide about the same
protection as the present standard at a fraction of the cost. The
dust mask alternative has been ruled out in the past because of
OSHA's commitment to the engineering controls approach to
regulation. In the case of cotton dust, this single-minded policy
generates no additional health benefits but imposes several times
the cost of an alternative policy relying on protective equipment.

Noise regulation exhibits even more dramatic variations in
cost-effectiveness. Table 7.4 summarizes the cost per worker protected from hearing loss in seventeen different industries.[11] The
cost variation is quite dramatic. Under the 90-decibel standard,
the cost per case of hearing loss prevented ranges from $19,000
for electrical equipment and supplies to $233,000 for the ma-
chinery industry. The disparity is even greater for an 85-decibel

TABLE 7.4. INDUSTRY COST VARIATIONS FOR THE OSHA NOISE STANDARD

Industry	Cost per worker protected (thousands of dollars)	
	90 decibels	80 decibels
Electrical equipment and supplies	19	39
Rubber and plastics products	38	68
Stone, clay, and glass products	53	96
Paper and allied products	62	78
Food and kindred products	75	179
Chemicals and allied products	80	132
Transportation equipment	87	111
Tobacco manufactures	104	200
Printing and publishing	108	215
Electric, gas, and sanitary services	137	189
Furniture and fixtures	150	151
Fabricated metal products	192	188
Petroleum and coal products	215	257
Primary metal industries	218	372
Textile mill products	227	395
Lumber and wood products	228	303
Machinery, except electrical	233	245
Weighted average	119	169

Source: Calculations by the author using data from Morrall (1976b).

standard, ranging from $39,000 to a high of $395,000 per case in the textile mill products industry. The per case estimates in the most costly instances are sufficiently high to warrant a major reassessment of whether it is worthwhile to spend hundreds of thousands of dollars to prevent a case of 25-decibel hearing threshold loss after twenty years of exposure.

The optimal noise standard that would emerge after one established some value per case of hearing loss would require the use of differential standards. Some industries, such as electrical equipment and supplies, should be tightly regulated; others, such as the chemicals and allied products industry, might be regulated at modest levels; in many instances in which the costs of noise reduction are extremely high, as for machinery and textile mills, no engineering controls may be warranted.

PERFORMANCE VERSUS DESIGN STANDARDS

The efficacy of a particular standard depends not only on its stringency but also upon the nature of the standard. OSHA safety standards typically specify precise characteristics that the workplace must meet, while health standards are defined in terms of objective workplace conditions rather than in terms of the level of risk. Standards pertaining to workplace characteristics are typically referred to as design standards, engineering standards, or specification standards. Standards stated more directly in terms of the reduction in worker risk are termed performance standards.[12] There is clearly a continuum of possibilities, depending on the leeway permitted by the regulation.

Specification standards can be found throughout OSHA regulations. Handrails are not simply required to be sturdy; OSHA instead meticulously defines their height (30 to 34 inches), thickness (at least 2 inches for hardwood and 1½ inches for metal pipe), spacing of posts (not more than 8 feet), and clearance with respect to the wall or any other object (minimum of 3 inches).[13] In the case of lawnmowers used professionally, OSHA not only imposes detailed requirements for lawnmower design but also insists on an overwhelming number of "caution" signs near each discharge opening, at the engine starting control point, near the opening for the catcher assembly, and in the instruction manual.[14] The most extreme case to date of engineering standards is that of the U.S. Department of Transportation's standards for city buses, which are almost tantamount to complete bus designs. (It is also noteworthy that most of these buses have developed major structural defects.)

Performance standards place a greater direct emphasis on the desired policy outcomes. The best-known performance standard now in existence is the EPA "bubble policy," which relaxes the pollution standards for each emissions point and imposes instead an overall requirement on emissions leaving a hypothetical bubble over the plant. Firms are then free to choose which pollution sources to control, thus reducing their compliance costs.

President Ford's Task Force on OSHA, headed by Paul Mac-Avoy, explored in detail the implications of such performance standards for machinery and machinery guarding. Current OSHA specification standards are so narrowly defined that they pertain to only 15 percent of all machines, and in these cases enterprise discretion is limited.[15] The model standard developed by the Task Force outlines the broad types of risks that may arise and the various options available to reduce these hazards, permitting the firm to select the least costly alternative.

In the extreme case of a job risk performance standard, there might be no regulations whatever regarding the safety of particular workplace conditions. If worker injuries and illnesses can be monitored accurately and if these outcomes are a reliable index of the risk facing workers, as they should be in large firms, a penalty could be levied on the firm based on the adverse effects of conditions on workers rather than on workplace characteristics.

Although performance standards may in many cases be superior to specification standards and should always be included among the policy options, it is an oversimplification to claim that they are always preferable. The choice between the two types of standards hinges on four types of issues. The first concern is the degree of certainty regarding the practical implications of the standard, which affects not only its equity but also the extent to which firms can make investments that they can be confident are in compliance. Engineering standards are specified quite precisely, and once these are met, there is no risk of penalty unless the regulation is changed.[16] If enterprises instead are exhorted to make the workplace safe through performance standards, the result may be capricious enforcement creating so much uncertainty that firms will forgo making improvements until after they are inspected and penalized.

How reliably one can predict the outcome under a regulation

is a quite legitimate policy concern. Insofar as possible, performance standards should be formulated in terms of objective criteria (lead levels in workers' blood, number of workers killed) to
provide a sound basis on which firms can make decisions. Even if
it is not possible to equal the predictability of a specification
standard, performance standards may have other benefits that
offset these differences.

A second oft-cited advantage of specification standards is that
they provide information to firms on how to improve safety.
Since most OSHA standards were prepared from voluntary industry standards, modified by changing the discretionary
"should" to a more demanding "shall," it is difficult to maintain
that OSHA did much more than make this information more
prominent. Health exposure standards set at the lowest technically feasible levels convey no useful risk information at all.
More generally, performance standards are not inherently incompatible with efforts to provide information about alternative
methods of compliance.

A third concern is which type of standard is more effective in
promoting health and safety. Under design standards, we do this
indirectly by penalizing workplace conditions, such as a slippery
staircase, which are believed to pose risks. In some cases the relationship between the environmental condition and the level of
risk is unclear. Cotton dust exposure levels are strongly but imperfectly correlated with byssinosis, and no causal link between
the two has yet been established. If cotton dust actually causes
byssinosis or if the reduction of dust also leads to a reduction in
the agent that does cause the disease, the cotton dust exposure
standard will be effective. It is also possible that the regulation
will not improve worker health even if cotton dust exposures are
reduced. In instances where we regulate workplace characteristics correlated with the risk rather than the risk itself, it is especially important to monitor the policy's effectiveness. For performance standards that are more clearly linked to health
outcomes than to the work environment, these problems are diminished.

The central advantage of performance standards is that the

firm has the opportunity to select the least costly means of compliance. The cost savings do not stem solely from the fact that businessmen have greater technical expertise than government officials, though this may be a pertinent factor. The greatest gains from this discretion arise from the wide variations in technologies of different vintage and type. Although one compliance approach may be most efficient in many contexts, uniform risk reduction technologies will seldom be optimal in all situations.

The scope of possibilities introduced by performance standards encompasses much more than simply offering a firm several alternative ways of guarding a punch press. Other, more imaginative responses can be introduced. Firms might require workers to wear respirators to decrease health hazard exposures rather than make more costly changes in the workplace. In the case of the noise standard, protective devices could achieve a 90-decibel exposure level at the cost of $15,000 per worker, as compared with $119,000 per worker for the engineering standards proposed by OSHA.[17] Protective devices of these types may, however, raise other problems. If they are uncomfortable, workers may not wear them and, if they do, they may require higher wages to compensate for the increased discomfort. Earmuffffs and other devices to muffle noise may prevent hearing loss, but they may also increase the risk of accidents if workers cannot hear warning shouts. These caveats suggest that the attractiveness of alternative modes of regulation should be assessed on a case-by-case basis. No single regulatory mode is likely to be always dominant.

A final possibility is that the enterprise may choose to meet a performance standard not by altering the workplace or protecting the worker but by rotating the workers exposed to the risk. The early, reversible stages of byssinosis, for example, are indicated by mild symptoms, so workers with these symptoms could be removed from their jobs before the disease reached an advanced stage. Similarly, workers exposed to lead could be rotated to different jobs once the lead in their blood reached levels linked to serious health effects. Cleanup operations in nuclear power plants are now routinely delegated to temporary workers to avoid exposing any single worker to a particularly large dose of radiation. However, since inexperienced employees are less efficient at these jobs, their exposure time will be greater, so the overall incidence of radiation-induced aliments may increase, depending on the relation between radiation exposure and adverse health effects.

Whether or not it is desirable to rotate workers depends on the nature of the health impacts and the dose-response relationships. If the early stages of byssinosis are sufficiently less harmful than the later stages, rotating workers will reduce the overall adverse health effects. When the rotation is based on some risk exposure, as in the lead case, the nature of the dose-response relationship assumes critical importance. If, for example, the risk of cancer rises linearly with one's exposure to radiation, on average no lives will be saved through rotation; but if there is a no-risk threshold before such a linear increase in the risk begins, there may be considerable health benefits from a rotation policy. As our understanding of the underlying determinants of worker health are improved, we will be better able to identify situations in which rotating workers is desirable.

PERSONAL CHARACTERISTICS AND OPTIMAL STANDARDS

The risk posed by a particular job varies widely for different individuals, as indicated by differing sensitivities to toxic substances or the greater risk of assault faced by short policemen. Many particularly controversial instances involve risks strongly correlated with one's sex or race. Blacks with the gene for sickle-cell anemia may incur a greater risk of harm from the low-oxygen conditions faced by a pilot, and female mail sorters have a greater frequency of back injuries when moving the standard seventy-pound mail sacks.

In the absence of government intervention, market allocations of individuals to jobs will promote efficient matchups in many instances. If the worker bears all of the harm associated with the risk and if he is cognizant of his own particular risk, not simply the average risk for all, he will select his job optimally. Since these informational requirements are clearly more stringent than when there are no major differences among individuals in the riskiness of a job, there may be an additional motivation for government intervention if the heterogeneity of the risk is not well understood.

The employer and the worker's fellow employees also may have a stake in the allocation of workers to jobs. Worker injuries and illnesses disrupt production, lead to additional training costs, boost workers' compensation benefits, and affect the firm's reputation, which in turn alters wage rates. A worker's careless behavior may also result in injuries to other workers.[18]

For many jobs involving strength, dexterity, and other risk-related physical characteristics, employers are very selective in filling positions. Sometimes this selectivity is related quite explicitly to the risk. Smokers are not permitted to work at the Johns-Mansville asbestos plant because they face a risk of lung cancer almost a hundred times greater than nonsmokers. Similarly, no women are permitted to work in the pigment paint division of the American Cyanamid Corporation because lead exposures pose considerably larger risks to pregnant women. Distinctions based on sex have been widely condemned, in part because the women previously working in the division agreed to become sterilized to keep their jobs. In contrast, the ban on smoking received widespread favorable publicity and, unlike the lead case, did not lead to a critical OSHA review. Society may feel more strongly about distinctions based on unalterable personal characteristics than about individual choices that have increased one's riskiness in a job.

Another mechanism by which markets might reduce the costs imposed on others from this heterogeneity is by altering the worker's wage rate. Accident-prone workers who impose greater losses on others should be paid a lower wage to reflect these expected costs. This wage flexibility is often limited in practice by wage floors (such as the minimum wage), limitations on the variation of wage rates if the source of the heterogeneity in riskiness is highly correlated with personal characteristics (particularly race, sex, and age), and institutional rigidities that prevent variation of the wage structure on an individual basis.[19] Financial mechanisms also may have inherent limitations since there can be no adequate *ex post* compensation for a worker who is killed as a result of hazardous behavior by his coworkers.[20]

Once the costs are imposed on parties beyond the labor market transaction, market processes become even more inadequate. The imposition on taxpayers of social insurance costs for injured workers has traditionally been the most pressing concern of this type. Much more disturbing problems have arisen in the past few decades as we have begun to learn more about the possibly catastrophic implications of workplace exposures for fetuses subjected to radiation, lead, and other carcinogens. Although the mother may take the baby's interests into account in selecting a job, there is no assurance that the preferences of the unborn will be fully reflected in her decision. Moreover, the mother may not have complete information regarding the risk to the fetus, or she may be unable to alter the risk. For example, a woman with high lead levels in her blood before becoming pregnant will continue to have possibly hazardous lead levels even if she leaves the job associated with the exposure.

Situations such as these may pose major difficulties for the employer as well. If he prevents such exposures by excluding all women from the job, he may exclude a substantial number of workers who would not have had a baby exposed to the risk. Alternatively, failure to discriminate in this fashion may increase his liability and the pressure to incur the costs of preventing possible adverse outcomes. In such a case government regulations that define the employer's obligations precisely may benefit the employer by sharing some of the responsibility for his decisions.

To the extent that these issues have been addressed, it has been through rather simplistic policies. Most efforts have attempted to eliminate the risk for all, irrespective of the cost. While the past performance of affirmative action programs indicates an apparent willingness to sacrifice economic efficiency to promote other social objectives, this willingness is not unbounded.

Rather than adopt rigid policies that mandate, somewhat implausibly, the objective of equal risk for all, it would be preferable to review the need for regulations to protect high-risk groups on a case-by-case basis. If the cost of reducing the risk is small compared to the benefits to these groups of increased access to jobs, such measures should be pursued.[21]

But in any case, some heterogeneity in riskiness will remain. We cannot provide jobs of equal safety for everyone any more easily than we can ensure that all individuals will be productive on a particular job irrespective of their strength, diligence, or intelligence. Indeed, attempts to promote such equalization undermine a major beneficial feature of all market allocations. Workers are not assigned to jobs at random throughout the economy, because the economy functions much more effectively if the differences in individual productivity on different jobs are reflected in the job matchups. An individual's sensitivity to various hazards or proclivity toward accidents is simply one aspect of his overall productivity. If this heterogeneity is exploited in matching up workers to jobs, rather than suppressed, the overall safety and efficiency of the economy will be enhanced.

8

THE POLITICAL CONTEXT OF RISK
REGULATION POLICIES

THE DISMAL PERFORMANCE OF OSHA IS BY NO MEANS AN ABERRA-
tion. Most other risk regulation agencies have come under similar, if somewhat less harsh, attacks. A consensus appears to be
rapidly emerging that the modes of regulation have not been well
chosen and that the design and administration of these regulatory strategies have been inadequate and often simply inept.
More specifically, the risk regulation agencies share three principal shortcomings: 1) failure to consider the economic basis for
intervention, with reliance instead on a risk-based criterion, 2)
exclusion of cost-risk tradeoffs except through very crude and
often ad hoc cost considerations, and 3) reliance on rigid standards, which tend to be engineering controls.

Neglect of the economic rationale for intervention is certainly
the most pronounced common deficiency. There are, however,
some minor differences in how agencies justify and design their
regulations. For example, whereas OSHA now focuses on risks
judged to be "significant," the Food and Drug Administration's
drug regulations focus on "substantial" risks, the Consumer
Product Safety Commission's (CPSC) efforts address "unreasonable" risks, and food safety requirements under the Delaney
Amendment attempt to ensure zero carcinogenic risk.

What the agencies mean by risk also differs. OSHA's regulatory analyses have been the most sound, since they typically address the annual risk per full-time employee. In contrast, the
Environmental Protection Agency's (EPA) air quality stan-

dards focus on a "margin of safety." As a consequence, its lead standard and petrochemical oxide standard were based on the health effects for only the most sensitive members of the population, ignoring both the number of people not affected and the level of the risk probability for those who are sensitive to the exposure. Indeed, for the lead standard, EPA required that 99.5 percent of this most sensitive group (primarily children) be exposed to lead levels below a zero risk threshold.

The CPSC is perhaps the most inconsistent, because in justifying some regulations it counts the total number of injuries, while for others it focuses on the annual frequency of injuries for users of the product. In each case, there is no adjustment for the extent of a product's use by the consumer. Moreover, the levels of the risks targeted for regulation are often quite small—death risks on the order of 1 in 100,000 annually—so the agency may be banning products that are safer than those that will take their place.

The exclusion of cost-risk tradeoffs also differs among agencies since cost considerations usually do enter, at least in part. OSHA is typically concerned with the number of plant closures, while the CPSC usually avoids banning products of clear value to consumers but which pose safety hazards that are difficult to influence through standards. EPA has the most elaborate approach. Past debates over water pollution standards have focused on whether firms should be required to adopt the best available, best conventional, or best practicable pollutant control technology. Although these categories represent decreasing degrees of stringency, their precise implications for policy are unclear, except perhaps to those privy to the inner workings of EPA. Cost considerations enter in a variety of ways, but the fundamental problem is that the cost-risk tradeoffs are not integrated into the policy design.

Finally, each of these agencies has relied on a rigid standards approach to regulation. Although there are some notable exceptions to the emphasis on engineering controls, for the most part the regulations provide little flexibility to firms attempting to comply with the standards. Indeed, other agencies' efforts are often in the spirit of OSHA's specification standards. OSHA's lawnmower standard, which mandates a variety of "caution" signs, has been followed by a CPSC lawnmower standard requiring "danger" signs and imposing extensive design requirements. The CPSC's twenty-four pages of bicycle safety standards, culminating with a tag testifying to the bicycle's safety, may even rival some of OSHA's now-legendary standards.

Widespread dissatisfaction with the efforts of risk regulation agencies is not new. The difficulty for policymakers has been to find a mechanism to channel these regulatory activities in more productive directions. There are three principal types of leverage that can be used to improve the regulatory process.[1] First, the legislative mandate can be altered either through changes in legislation or through the interpretation of this legislation by the courts. Second, a system of administrative review within the executive branch can be established. And finally, the officials appointed to these agencies can influence the design and administration of regulatory policies.

THE REGULATORY REVIEW PROCESS

Most of the emphasis in recent years has been on administrative review. Major changes in legislation or in the composition of the courts occur relatively slowly. Indeed, for most presidential administrations, the time horizons are sufficiently short that the legislation authorizing regulatory activities and the court's interpretation of this legislation are viewed more as constraints than as policy parameters to be manipulated. In contrast, the president can restructure the executive branch's oversight process for regulation fairly easily. Until the mid-1970s, regulations were reviewed relatively infrequently, and in most instances the reviews were not based on a thorough analysis of the regulations' effects.

Earlier in that decade, however, there had been a proliferation of new social regulation agencies, including OSHA, EPA, CPSC, and the National Highway Traffic Safety Administration (NHTSA). Their legislative mandates tended to be absolute, requiring the agency to promote health and safety without any explicit reference to cost considerations. The chief matters of concern were the technical feasibility of various alternatives and the impact of the policies on the agency's risk-based mandate.

In the absence of any internal checks within these regulatory agencies, President Ford established a framework to try to force regulations to be structured in a more efficient manner. Through Executive Order 11821 (November 24, 1974), he required that all major regulatory initiatives be accompanied by an inflationary impact statement. The threshold requirement for the statement was a cost impact from the regulation of $100 million in one year or $150 million over two years. Any proposed regulation above this threshold had to include an analysis of its effect on product supplies, competition, productivity, and the costs imposed on consumers and businesses. The mechanism for this statement was detailed in the Office of Management Circular No. A-107 (January 28, 1975), which added the requirements that benefits and costs of the regulation be compared and that the agency review policy alternatives.

A new agency, the Council on Wage and Price Stability (CWPS), also was established (November 1974) within the Executive Office of the President to oversee the impact statement process. More specifically, Section 3(a) of the act establishing the council gave it the authority to "intervene and otherwise participate on its own behalf in rulemaking, ratemaking, licensing and other proceedings before any of the departments and agencies of the United States, in order to present its views as to the inflationary impact that might result from the possible outcomes of such proceedings." As a consequence, CWPS was able not only to discuss proposed regulations with the regulatory agencies but also to make public its assessment of them. These assessments, usually in the form of critical evaluations of the impact statements, were filed with the agency during the public comment period. The extent of this authority was quite broad, as CWPS could comment on the activities of independent regulatory commissions, such as the Interstate Commerce Commission, as well as on proposals by executive branch agencies.

The regulatory oversight process remained largely unchanged

through the first year of the Carter administration, as inflationary impact statements were replaced by economic impact statements. Through Executive Order No. 12044 (March 24, 1978), President Carter strengthened the review process in two ways. First, agencies had to show that "alternative approaches have been considered and the least burdensome of the acceptable alternatives have been chosen." Second, Carter established the Regulatory Analysis Review Group (RARG). This group, chaired by a member of the Council of Economic Advisers (CEA), consisted of representatives from the White House (Domestic Policy Staff, Office of Management and Budget, and CEA) and from executive branch agencies, who served on a rotating basis.

The targets for RARG interventions were selected on the basis of the overall desirability of the proposals and the likely effect of a sound critique on the policy outcome or on the precedential implications of the policy. For up to ten of the most important regulatory filings annually, CWPS submitted a RARG analysis in lieu of its own assessment. CWPS continued to play a central role since its legislation provided the authority to make public filings, which other parts of the Executive Office of the President lacked. The RARG analyses were written principally by the staffs of CWPS and CEA and had to be approved by the RARG Executive Committee, so they represented the consensus view within the Executive Office of the President.

The regulatory monitoring by the Ford and Carter administrations was notably similar. The staff responsible for the regulatory analyses remained largely unchanged, as did the substance of the analyses and the nature of the conflicts between the White House and the regulatory agencies. The principal difference was the establishment of RARG, which may have increased the effectiveness of the review process somewhat.

A SUMMARY OF RISK REGULATION INITIATIVES

Table 8.1 is a numerical summary of the CWPS analyses from 1975 to 1980, including those submitted on behalf of RARG. The number of filings per year is listed for each agency whose activities can be broadly characterized as focusing on risk regulation. There was an average of fifty filings per year, of which one-third generally pertained to health and safety regulation.

Numerically, OSHA risk regulations constitute only a small portion of the total—fourteen of the ninety-six risk filings and only one filing in 1979 or 1980. In contrast, almost half of all filings and the preponderance of all analyses during 1979 and 1980 were for EPA. Despite the flurry of reaction to EPA proposals in these two years, the portion of CWPS analyses devoted to risk-related issues declined after the initial review period. For risk-regulation agencies other than EPA, the number of CWPS filings plummeted from thirty-eight in 1975–1977 to nine during the 1978–1980 period.

TABLE 8.1. PUBLIC SECTOR RELEASES ISSUED BY THE COUNCIL ON WAGE AND PRICE STABILITY REGARDING RISK REGULATIONS, 1975–1980

Agency	1975	1976	1977	1978	1979	1980	Total
CPSC	1	3	1	0	0	0	5
EPA	7	4	4	5	14	15	49
FDA	6	4	2	0	1	1	14
NHTSA	4	1	2	1	0	2	10
OSHA	2	4	5	2	0	1	14
Other risk agencies	1	2	0	0	0	1	4
Total risk filings	21	18	14	8	15	20	96
Total filings	52	53	44	34	51	66	300
Risk share of total	.40	.34	.32	.24	.29	.30	.32

Source: Tabulations by the author using CWPS files.

This falloff in regulatory activity was due in part to the agencies' uncertainty regarding the court's interpretation of their legislative mandate. Most important was the OSHA benzene case, which was viewed as a major court test of the criteria for carcinogen regulations and in particular of whether cost considerations must enter the standard-setting process.

Although there has been a noticeable decline in new risk-related regulations, the summary of CWPS filings alone does not provide a comprehensive index of existing regulations and the current impact of these regulations on the economy. Most important is that only selected major new regulations proposed since 1975 have been reviewed. Almost all of OSHA's safety standards were promulgated before any review process was established. In some instances, agencies may undertake new regulatory actions under their existing authority. For example, the CPSC has broad authority to ban or recall products posing substantial risk so long as it believes that doing so is in the public interest. The exclusion of these effects makes the tally in Table 8.1 more a measure of new regulatory authority than of the total regulatory burden.

To accurately assess changes in the extensiveness of this authority, one would like to know not only the number of major risk regulations but also their economic impact. A more instructive index is the level of costs associated with the regulations. Table 8.2 summarizes the regulatory costs for all agencies engaged in the regulation of health and safety risks, expressed as the discounted present value of costs associated with regulations proposed in any year rather than the actual costs incurred within particular years.

The cost summary is very comprehensive, as only a few minor filings lacked cost data. A range of cost levels is given in instances in which the agency discussed more than one regulatory alternative. For each regulatory policy I utilized only the most accurate cost estimate (usually that of CWPS).

The total costs implied by the risk-related regulatory proposals in the 1975–1980 period is staggering—from $300 billion to over $800 billion, or an implied annual cost of $30 to $80 billion. The

TABLE 8.2. Costs (billions of dollars) of proposed risk regulations reviewed by the Council on Wage and Price Stability, 1975–1980

Agency	1975	1976	1977	1978	1979	1980	Total
CPSC	—	1.1–1.8	2.3–3.4	—	—	—	3.4–5.2
EPA	8.5–26.2	6.1–6.5	0.8	148.1–196.4	27.7–37.1	26.4–29.4	217.7–296.4
FDA	0.6	0.02	1.2	—	0.90–4.0	0.3–0.4	3.0–6.2
NHTSA	0.7–1.4	—	—	2.8	—	11.1–23.5	14.5–45.6
OSHA	17.3	2.7	4.4–24.1	69.2–448.2	—	—	93.6–492.3
Other	0.1–0.8	—	—	—	—	0.02	0.1–0.8
Total	27.2–46.3	9.9–11.0	8.7–29.5	220.1–665.3	28.6–41.1	37.8–53.3	332.3–846.5
OSHA percentage	37%	25%	51%	31%	0%	0%	28–58%

Source: Calculations by the author using CWPS filings and supporting data. All dollar figures are in terms of the prices for the particular year. In calculating these present values, I used an interest rate of 10 percent. If these costs had been expressed instead on an annualized basis to reflect the implied level of annual costs when spread over an infinite period, the cost figures would have been scaled down by a factor of 10.

estimates of the cost of proposed regulations are a reasonable
index of actual costs, since the policies adopted closely parallel
these proposed efforts. In the discussion below, I will focus not
on annualized costs of these policies but on the present value of
the costs to emphasize the long-term commitment of resources
they represent.

The breadth of the cost range is due almost entirely to the 1978
OSHA carcinogen policy, whose costs were projected to range
from $69 billion to $448 billion. Unlike most proposed rulemakings, the proposal specified no particular standards but rather
general principles to govern carcinogen regulation, leading to
widely divergent cost estimates that hinged on the degree of
stringency of the guidelines. For the intermediate case, termed
the "medium scenario" in the economic impact statement, the
present value of costs was $247 billion, approximately the middle
of the cost range.[2]

The potentially substantial costs associated with the carcinogen policy contribute to OSHA's major role—28–58 percent—in
risk regulation costs for 1975–1980. As the bottom row of Table
8.2 indicates, from 1975 to 1977 OSHA costs played a similarly
major role even without the costly carcinogen policies, which account for almost all of OSHA's 1978 costs. During 1979 and
1980, however, new OSHA regulations ceased to be a factor since
the uncertain judicial interpretation of the agency's mandate led
to a moratorium on major regulations.

THE IMPACT OF THE REGULATORY OVERSIGHT PROCESS

While Tables 8.1 and 8.2 indicate that CWPS was active in analyzing a large number of costly regulations, they do not imply
that these efforts were effective in reducing the regulatory burden. Before assessing the impact of these analyses, it is helpful to
begin by asking what one could reasonably expect from this kind
of regulatory review structure.

For major regulations, agencies were simply required to pre-
pare cost estimates for alternative policies, which CWPS then
reviewed and discussed in its public filings, sometimes in con-

junction with other groups in the executive branch. Agencies were not required to meet any test for proposed regulations except preparation of an economic impact analysis. The review by CWPS was a public comment on proposed regulations; it could not veto a new regulation.

Some critics of the review process claim that it has failed because some bad regulations continued to be promulgated, and more generally, because the nature of social regulation was not radically transformed. However, it seems inappropriate to judge the review agency's performance by standards it could not have met, given its limited authority. To the extent that there has been progress, it has been through slow evolution rather than any stark change.

To properly evaluate the effectiveness of the review process, ideally one would like to compare the regulations actually issued with those that would have been promulgated in the absence of these procedures. Reliable information of this type is not available even on an anecdotal basis, since few agency officials are willing to discuss the extent to which they have been affected by the oversight process.

It is, however, possible to cite several important advances. First, and perhaps most fundamental, is that agencies now calculate the costs of proposed regulations and, to a somewhat lesser extent, their benefits. When benefit estimates are provided, they are typically expressed in physical units rather than monetary terms. The estimates are generally believed to be usually unbiased, and when it is clear that a conceptual error has been made, the supporting data are often available for making an analytically correct estimate.

While more information about the economic implications does not ensure efficient regulation, it does provide a basis for more substantive debates concerning appropriate policy. By reducing the regulatory agency's almost exclusive control of the data pertaining to a regulation's effect and by promoting the generation of data pertinent to responsible policy evaluation, the review process serves to strengthen efforts within and outside regulatory agencies to make regulations more efficient.

In some instances, debate is fostered by the review process itself. CWPS filings usually required frequent contacts between CWPS analysts and their regulatory agency counterparts, as well as periodic briefings on prospective policies. In the case of RARG interventions, the RARG Executive Committee periodically met with representatives from the affected agencies to discuss proposals that might be analyzed and, when a RARG intervention was warranted, there were several meetings to debate the substance of the filing. Regulatory agencies were typically represented in these meetings by economists from offices that were established or expanded after the advent of the regulatory oversight process. An important dividend of this impact statement and the review process in general is that it has led to an expanded role for economic analysis within regulatory agencies.

Although the overall effect on regulation is difficult to assess, an important change that can be reasonably attributed to the oversight effort is that now the economic consequences of regulations are a central part of most policy debates. These concerns were seldom raised in agency discussions before the review process was in effect. Also, the majority of regulations undergo noticeable improvement in the period between the initial proposal and the final regulation. In almost all instances of RARG interventions, the final regulation was closer to the RARG position than was the initial proposal.

The exact source of this improvement is unclear. The RARG interventions typically focused on the most important regulatory interventions, benefited from extensive and insightful CWPS-RARG analysis, and received the greatest amount of White House followup by the CEA, the Domestic Policy Staff, the Office of Management and Budget (OMB), and the inflation advisor to the president, Alfred E. Kahn. These factors are strongly interrelated. White House followup relies on information in the public record, such as material in the CWPS-RARG analyses. The subsequent political impact of sound analyses provides an incentive for producing effective regulatory analyses, which in turn enhances their eventual usefulness to those who will rely on them in their efforts to redirect regulatory activities.

While one cannot readily disentangle which aspect of intervention is of most consequence, what is important is the net effect of the regulatory oversight process. It is not accurate to claim, as have some critics, that only the most ill-conceived regulations were eliminated. For most regulations there was some appreciable modification that enhanced their economic merits. However, only in the case of rate regulation by agencies such as the Interstate Commerce Commission has there been a dramatic change in the nature of regulatory proposals.

The progress that was observed in the revision of regulations may understate the ultimate impact of the analyses. Even when the CWPS-RARG recommendations were not adopted, they did provide the informational base for future regulatory reforms and sunset actions, should there be a change in the political climate for regulatory reform. In 1980, for example, the Carter administration undertook an economic aid program for the auto and steel industries, which included a substantial regulatory relief component. The agenda of regulations that were targets for modification was based largely on the findings of the earlier analyses. Moreover, the CWPS and CEA staff that had developed expertise in these issues as part of the review process were able to provide primary staff support for the regulatory overhaul that arose out of Carter's industrial policies. The flurry of sunset actions at the start of the Reagan administration and the sunset targets selected in 1981 and 1982 also were based almost entirely on past analyses by the CWPS staff.

THE LIMITS OF REGULATORY OVERSIGHT

Notwithstanding the beneficial aspects of regulatory oversight, serious deficiencies remained in the majority of regulations promulgated. The most fundamental inadequacy in the CWPS-RARG framework for oversight was the lack of effective political power by the proponents of sound regulation. The CWPS-RARG analyses were not binding in any way and at best only served to bolster efforts to rationalize the regulatory process. Even if there was a consensus within the Executive Office of the President that a regulatory proposal was flawed, this opposition might be unsuccessful unless the president's support was also elicited.

The large number and technical complexity of regulations make it infeasible for presidential intervention to be more than an occasional factor. When a regulation is appealed to the president, there is no assurance that the economic merits of the issue will dictate his decision. In the widely publicized case of the OSHA cotton dust standard, the CEA chairman brought the issues raised by the CWPS analysis to President Carter, who initially opposed the tight standard then later reversed his position. In supporting the position of OSHA, Carter appears to have been influenced more by the prospect of alienating organized labor than by the consensus economic views within the White House. In a similar instance, Carter supported the concept of industry liability for Superfund legislation, thus concurring with the EPA position, which had been attacked by economics-oriented White House groups, such as CWPS, on the grounds that the penalty was not linked to the generation of the hazardous wastes.

Since the impact of policy decisions on the president is not influenced only by the net economic effects of regulations but also by political factors, such as whether he is perceived as a good environmentalist or as being loyal to organized labor or business, these outcomes reflect the incentives facing most presidents. Moreover, when regulatory policy issues are brought to the president at the final stages of the regulatory process, often after several years of debate and analysis, the political positions of most affected groups are well established and cannot be readily altered by presidential persuasion.[3]

The legislative mandates of the regulatory agencies also impede the effectiveness of efforts to introduce consideration of the costs of regulation into policy formation. In extreme instances, such as the ambient air quality standards under the Clean Air Act, EPA is prohibited from considering costs. Much more prevalent is the practice of agency officials retreating behind supposedly absolute legislative mandates to promote health, safety, or environmental quality. Since the degree to which agencies *may* consider costs and the degree to which they *must* consider costs both have not been fully resolved by the courts, the internal debates over the requirements of the legislation are seldom productive. Indeed, when policy discussions begin to focus on these legal issues, it is usually a signal that constructive discussions have broken down and that the agency is not going to be swayed by the implications of any further economic analysis. A revision of these legislative mandates to introduce a benefit-cost or cost-effectiveness requirement is essential if substantial reform is to take place.

Insofar as the oversight process has an impact on the regulations that an agency can issue, it will also impinge on the design and enforcement of the regulatory policy. Certainly in the early stages of CWPS, the inflationary impact statements were undertaken after the fact, as agencies such as OSHA contracted for outside studies to provide the necessary supporting analysis for their policies. The subsequent establishment of economic analysis units within OSHA and other agencies has reduced somewhat the classic problem of economic analyses being used to justify decisions rather than to guide decisionmaking. If regulatory analyses are ever to serve much more than a subsidiary function, the analysis staffs within OSHA and the other agencies should be greatly expanded and given a considerably larger role in policymaking.

There is a general consensus among supporters of regulatory reform that the review function required strengthening in some manner. One possibility is to introduce presidential or congressional review of regulatory proposals. As past experience has indicated, neither of these institutions is particularly well suited to examining a large number of regulations and the often complex arguments regarding the merits of alternative policies. Even if the issue can be thoroughly explored, there is no assurance that the decision will be motivated by the net social benefits of the regulation rather than by quite different political pressures.

This difficulty is endemic to any kind of reform that relies on review at the final stages of policy formation. To be successful, the criteria for sound regulations must be integrated into the early stages of policy design. Otherwise, analysis and review tend to be a minor adjunct to the policymaking process, in part because of political pressures that become established once policies begin to evolve. While these forces do not always support the proposed policy, they invariably create a more sharply defined group of interests in which the review process can operate.

REAGAN'S REGULATORY OVERSIGHT PROCESS

Shortly after taking office, President Reagan abolished the Council on Wage and Price Stability and shifted its regulatory review staff to the Office of Management and Budget to utilize the political power wielded by OMB in the budget-making process.[4] With his Executive Order 11291, Reagan established a benefit-cost test for proposed regulations, a formal mechanism for designating existing regulations as candidates for sunset review, and a process by which a regulation must be approved by OMB before it is formally proposed. To reduce the number of conflicts that might be appealed to the president, Vice-President Bush was given broad authority over regulatory issues.

Despite the central role of regulatory reform in Reagan's economic program, the scale of the oversight group remained unchanged, with only about twenty people (including support staff) responsible for analyzing the economic merits of proposed regulations. Unless this policy analysis staff is expanded considerably, it will not be able to devote much attention to the evaluation of existing regulations and the establishment of targets for sunset reviews, much less to careful analyses of all prospective regulations.

Although it is too soon to assess the long-term impact of these changes, some preliminary conclusions can be drawn. The principal shortcoming of Reagan's reorganization of these functions is that he sacrificed the legal authority of CWPS to intervene publicly in rulemaking proceedings of independent agencies and executive branch agencies. This change affects both the scope and nature of the regulatory oversight process. Whereas CWPS formerly reviewed and commented upon the regulations of independent regulatory agencies, such as the Consumer Product Safety Commission, the White House no longer has the explicit legal authority to do so. The oversight of regulatory actions by executive branch agencies has also been limited, as OMB can review these regulations but cannot submit comments for the public record.

The shift to private interagency negotiations may make the motives for regulatory decisions suspect, since the basis for the actions will not be fully disclosed. More important than the perceived legitimacy of the process is the change in the nature of oversight. The CWPS public filing mechanism led the review staff to develop detailed, written analyses of the merits of proposed regulations. These analyses were the subject of often intense debates within CWPS and, in the case of RARG filings, were developed on a collaborative basis by the White House staff.

The new procedures do not have these built-in institutional mechanisms to ensure that a review is complete and sufficiently accurate to withstand public scrutiny. OMB analysts now may simply reject regulations on the basis of a cursory review unless the administrator of the process establishes an internal mechanism to promote thorough analyses. Even if such supporting analyses are prepared, these internal documents will not serve to stimulate public debate or to provide support for judicial actions against the regulatory agency. Moreover, since internal evaluations typically are not produced with the same care and do not have the same permanency as publicly released reports, there will be a lesser basis for selecting targets for future sunset actions.

The advantages of the Reagan approach stem from OMB's authority to review regulations before they are proposed and to require a formal benefit-cost analysis. In practice, this authority enables OMB to delay undesirable proposals indefinitely by rejecting the proposal's supporting analysis. The ability to halt new regulatory activity does not, however, greatly assist OMB in forcing agencies to reexamine existing regulations. OMB continues to have little leverage over sunset actions and has conse-

quently been much less successful in revamping regulatory policies than in decreasing the amount of new regulatory activity.

By increasing the degree of oversight authority and shifting
the review process to a period before the positions of the various
political actors become well established, the new oversight process should be better able to influence the final policy outcome.
This advantage may, however, be partially lost if agency heads
publicly commit themselves to propose or issue new regulations
by a particular date. This commitment is then used to gain leverage against the OMB oversight group, which must assume responsibility if the regulation is not forthcoming.

The Reagan reforms have also strengthened the substantive aspects of the oversight process by mandating a benefit-cost test.
This change has proven to be inconsequential for OSHA regulations, however, since the Supreme Court has prohibited the use
of a benefit-cost test for standards pertaining to toxic substances
and hazardous physical agents. OMB can, however, reject OSHA
proposals on other grounds if the supporting regulatory analysis
is inadequate. The continued problems raised by the OSHA legislation point up the need for fundamental legislative reform.

The Reagan oversight procedures have increased the institutional authority of the review group but have sacrificed much of
the scope of the CWPS operation and have severely limited public access to the discussions underlying regulatory decisions. The
net effect of this change should be to enhance the effectiveness of
the regulatory review process. As in the case of earlier efforts,
however, the impact of oversight will depend more on the president's ultimate commitment to regulatory reform than on the
specific structure of the oversight process.

REGULATORY BUDGETS

An alternative to presidential or congressional review is the establishment of a regulatory budget. Within this framework,
OMB would assign executive branch agencies a total budget for
the annual costs imposed on the economy by the agencies' regula-

tions. Agencies would then issue and enforce regulations subject to this constraint. The principal attraction of this approach is that it would impose a systematic form of discipline on regulatory agencies. Since the central budgetary institution is already in place and is relatively effective in controlling expenditure programs, this type of budgetary leverage has obvious political benefits. The Reagan oversight group is located within the OMB but does not explicitly use budget restraint as a mechanism for promoting sound policies.

While the budget concept certainly merits serious consideration, one should be cautious in ascribing to it all of the prospective benefits that have been claimed by some of its more enthusiastic advocates. First, the administrators of the regulatory budget cannot simply establish budgetary limits for an agency and then set in motion a decentralized process whereby the agency can choose the regulations that will exhaust its budget. If the inadequacies of regulation simply derived from their excessive degree, setting cost limits for each agency could potentially be an optimal strategy. But the problem is typically that society is saddled not only with too much regulation but also with the wrong kinds. Budgetary limits will not ensure the use of performance standards rather than design standards or the efficient enforcement of regulations.

Somewhat surprisingly, a decentralized process under a regulatory budget is sometimes touted on the basis that regulatory agencies have different objectives from those of a centralized agency such as CWPS or OMB and that they should be free to pursue these objectives. If anything, the differences that exist support the need for greater centralization in an agency that can focus on broadly defined benefits and costs rather than narrowly perceived objectives within a regulatory agency's parochial interests.

A regulatory budget will also encounter nontrivial implementation problems. Any analogy with budgetary processes for expenditure programs conveys a degree of precision that is largely illusory. The cost estimates for most regulations are necessarily imprecise since the impacts are often diffuse, and the implications for different industries are not well understood. A principal difference between regulatory budgets and expenditure budgets is the feedback provided on the accuracy of the estimates. For expenditures, the actual budgetary allocations each year are known, making it possible to identify biases in the cost assessments. There is no comparable accounting of the actual *ex post* costs of regulations. Certainly some monitoring mechanism can and should be established to assess these costs, but the accuracy of the measures of actual cost levels will never be as precise as expenditure cost data, nor will the cost of acquiring the information be as low.

Despite the focus on costs, benefit considerations also must enter. If budget levels are to be set correctly, one needs some indication of the benefits associated with different levels of expenditure. At present, benefit information is even less adequate than the cost data. When it is available, it is often in physical units: cases of hearing impairment prevented by the OSHA noise standard, number of burn victims prevented by a CPSC sofa flammability standard, or levels of lead concentrations in the blood of different population groups under an EPA lead standard. It is conceptually quite difficult to make comparisons across agencies or even within agencies with diverse activities affecting quite different aspects of individual welfare. In practice, the comparisons are complicated by political pressures to give priority to the most severe classes of risk or to those most prominently featured in the media, rather than those that can be reduced most effectively by government action.

For a more conventional regulatory review process, benefits must also be assessed. But this is usually done on a case-by-case basis, focusing on whether the benefits outweigh the costs. A regulatory budget is clearly more comprehensive in that it requires comparing the marginal benefits and costs of the activities of different agencies. The informational requirements of a regulatory budget are greater, as are the potential benefits to society if the budget functions effectively.

The principal advantage of the budget approach is that it offers a potentially effective method of gaining political control of the regulatory process. It does not in any way eliminate the need for scrutinizing the benefits and costs of regulatory alternatives.

Whether increased discipline of regulatory agencies can be better achieved through a regulatory budget or through a more stringent regulatory review process, such as a requirement that benefits be shown to exceed costs, is more of a political question than an economic one. Moreover, the advantages of these alternatives cannot be assessed in the abstract since several factors enter, such as the role of various agencies in different political administrations.[5]

As the regulatory oversight process becomes increasingly effective, the task for critics of risk regulation will also become more complex. As policies are placed on a sounder economic basis, it will no longer suffice to dismiss these efforts out of hand because of their complete neglect of cost considerations or some other fundamental error. Rather, we will need a much more precise notion of the proper role for regulatory policies and the manner in which they should be structured. The view I advocate in the final chapter is that this approach should be market-oriented and, in particular, based on the risk preferences reflected in worker choices.

9

Controlling Risks through Individual Choice

Many changes of a limited nature could be made that
would enhance OSHA's performance. Unfortunately, both the
design and enforcement of the agency's policies are so seriously
flawed that we must consider more extensive changes rather than
a minor tinkering with the levels of the standards or penalties.
The fundamental inadquacy of this and other risk regulation
agencies is that a basic mistrust and misunderstanding of market
forces pervade their operations. Nonzero risk levels are regarded
as evidence of market failure; standards are set at the most stringent technically feasible levels; and until recently the approach
to risk regulation reflected a general suspicion of the business
community and of economic analysis.

Toward an Effective OSHA Policy

The proposal I will sketch here is not intended to be a detailed
blueprint for reform. Rather, I will attempt to indicate the broad
direction OSHA policies should take. Underlying these recommendations is my belief that the agency has a potentially productive role, but that it should promote efficient levels of risk by
working in conjunction with market forces. Similar reforms
would enhance the performance of other risk regulation agencies
as well.

   I will discuss OSHA's future direction in terms of the desired
form of its policies, not how they should be packaged for the

greatest political salability. While the general thrust of these policies does stem quite directly from objective economic principles, a considerable element of personal judgment is necessarily involved in translating economic foundations into proposed policy frameworks. Alternative approaches also may be consistent with the types of economic concerns I have emphasized.

In many policy contexts, those who propose a shift in an agency's efforts choose to do so in quite muddled terms so that the resulting ambiguity will diminish potential opposition. Here I will take the opposite approach, making the intent of my proposal quite explicit. My sweeping proposals for reforming OSHA should not be regarded as a disguised attempt to dismantle the agency. The OSHA I envision will be much more aggressive than in the past, but the focus and spirit of its policies will be quite different.

After abolishing the current set of health and safety standards, we should base OSHA policies on the following three elements: (1) provision of risk information to workers, (2) greater merit rating of workers' compensation, and (3) penalties on hazardous firms to promote health and safety for selected risks. The greatest emphasis should be placed on the first of these components, which includes additional research on occupational health and provision of job risk information to workers. This will enable workers to make more efficient job choices that reflect their attitudes toward risk. These choices in turn will augment the already powerful economic forces of the marketplace, providing stronger incentives for safety than the present absolutist approach.

The desirability of an information-oriented approach derives in large part from the heterogeneity characterizing most markets. Workers differ in their attitudes toward risk, their susceptibility to various hazards, and their willingness to alter other activities, such as cigarette smoking, that affect the risks they face. Present OSHA standards abstract from this heterogeneity in an attempt to achieve a workplace free of significant hazards. Such an approach prevents individuals from accepting wage-risk combinations they find attractive and, where the standard does not completely eliminate the hazard, creates a situation in which

workers may incur the remaining risk unknowingly. As the
analysis of several OSHA standards indicated, the substantial
heterogeneity in the costs to employers provides an additional rationale for avoiding uniform standards. An effective risk information strategy will enable workers to be matched efficiently to
jobs so that the risks will fully reflect the values these workers
place on their well-being and the costs to employers of ameliorating the hazard.

Unfortunately, providing information does not necessarily ensure that workers have perfect information, nor is there always
an objective measure of risk that can be conveyed to workers and
effectively utilized in their decisionmaking. Our still-vast ignorance of many health risks and the difficulty in ascertaining the
risk associated with each job-worker combination limit any effort
of this type. Moreover, the link between information and its interpretation by workers as they make decisions is still not fully
understood.

A useful starting point might be to require employers to ap-
prise workers of the nature of the risks they face, the risk level of
the firm (death, injury, and illness rates), its relative risk compared to that of other firms in the industry, the possibility of
synergistic effects such as that of asbestos and cigarette smoking,
and the need for exercising special care. What should be avoided
is the carcinogen-of-the-week syndrome in which workers are
continually bombarded with warnings of yet another hazard, im-
peding their efforts to distinguish the relative riskiness of different choices.

Previous chemical labeling proposals perhaps best illustrate
the pitfalls to be avoided in an information-oriented approach.
In the waning days of the Carter administration, OSHA proposed chemical labeling standards that would have overwhelmed
workers with a detailed, polysyllabic list of the contents of every
possibly hazardous mixture.[1] While this information would have
been of potential relevance to a handful of experts in occupational medicine, the average worker would learn very little about
the risks he faced. Under the Reagan administration the emphasis has shifted toward performance standards, with labels
identifying the existence of a hazard and the nature of the risk

involved. The general spirit of this effort is consistent with the labeling system adopted voluntarily by the paint and coating industry. Such an approach is far preferable since the policy objective is to provide information that workers can understand. They can then make efficient job choices, distinguish the situations in which they should exercise greater caution, and seek prompt medical attention after exposure to potentially hazardous conditions.

Since we have little experience with information-oriented policies, except for rudimentary efforts such as warning labels on cigarette packages, OSHA should promote a diversity of approaches, monitor the outcomes, and attempt to identify particularly effective strategies. By providing some leeway for different firms to choose the most effective way to provide workers with risk information, it can promote this type of experimentation.

After these strategies are implemented, OSHA should attempt to ascertain the effect of information on workers' risk perceptions, on the turnover of present and future workers, and on the prevailing wage rates. If a warning of a very minor risk generates risk premiums reflecting an implicit value of life of $20 million or triples the firm's turnover rate, this form of information is probably unduly alarmist. This type of ongoing policy evaluation has been all but absent and should play a prominent role in framing OSHA policies.

The second component of the policy is to increase the merit rating of workers' compensation. While most suggested reforms of workers' compensation have focused on providing adequate benefits, a reform with potentially greater impact on worker welfare is to strengthen the link between a company's risk record and its insurance premiums. For the majority of firms, particularly small enterprises, there is presently little or no such relationship. Increased merit rating will eliminate the subsidies now conferred on unsafe firms and will provide greater incentives to reduce the types of accident risks that are the focus of the program. As I indicated in Chapter 5, such a change would not be tantamount to self-insurance and need not threaten the financial viability of small firms.

Additional incentives for safety could be provided by raising

premiums above the level needed to fund benefits. I am not advocating such an injury tax here because there is little evidence
that there is a significant inadequacy in the manner in which
market forces promote safety. The already powerful incentives
for safety created by the risk premium and turnover mechanisms
will be bolstered by the provision of risk information and the increased experience rating of workers' compensation. Until there
is strong evidence of a substantial market inadequacy in dealing
with these perceived risks, it makes little sense to institute a comparatively inconsequential injury tax. Such fine tuning will
have a negligible effect on safety and will only dissipate the impact of OSHA's more productive efforts.

The third component of the policy is intended to address the
instances in which a market process in conjunction with greater
risk information will not generate efficient risk levels. In these
exceptional instances, which should pertain almost exclusively to
dimly understood health risks, OSHA should consider additional
forms of intervention. If workers find it difficult to process risk
information pertaining to very low-probability risks, which often
have a deferred impact on their well-being, a more direct regulatory strategy should be considered. Justification for such intervention should be required on a case-by-case basis; the existence
of a workplace hazard should no longer be taken as a self-evident
reason for government regulation.

Rather than impose standards on such workplace conditions, it
would be preferable to impose penalties linked as directly as possible to the risk level. The penalty approach enables firms to
compare the benefits (as reflected in the penalties) and the costs
of needed changes on a decentralized basis, avoiding both the
need to set differential standards for each case and the social
costs associated with uniform standards. Ideally, the penalties
should be linked directly to workers' health. The objective is to
enhance workplace health and safety efficiently, not to generate
the intermediate outcome of altered workplace characteristics. In
many cases it is possible to monitor worker health directly, as
with byssinosis (lung function loss) and blood lead levels, so that
the regulatory incentives can be tied directly to the workers'

well-being. When it is not possible to monitor worker health accurately, a penalty system for hazardous occupational exposures could also be part of this system. The criteria for these penalties should be made as performance-oriented as possible but should be designed in a manner that does not create substantial uncertainties for firms.

Since enterprises will pay penalties for hazardous outcomes if they find it efficient not to correct a situation, the enforcement effort assumes a quite different character than it has when inspections are used to identify noncompliers who are then compelled into compliance by ever-escalating penalties, regardless of the cost of compliance. Indeed, a firm's compliance status is no longer a fundamental concern. Enterprises choosing a more hazardous work environment are not attempting to evade OSHA but are exercising their choice to not reduce the hazard because the costs of doing so outweigh the health benefits to workers, as reflected in the penalty schedule established by OSHA. The intent of OSHA policies will be to establish efficient levels of health and safety, not risk-free environments.

Once OSHA has shifted to this nonconfrontational approach, its enforcement effort could become much more comprehensive. For example, firms could self-assess their own penalty levels based on their risk performance. OSHA inspectors would assume a role akin to that of tax auditors, providing an incentive for firms to self-assess the penalties accurately. Inspections could then be selective and would not be required for the firm to be affected by the penalty structure. This system would give OSHA much more extensive coverage of the workforce than the present system, in which firms rarely see an OSHA inspector.

Critics of the approach I have suggested no doubt will maintain that the abolition of the present standards and this shift in OSHA policy will endanger workers' lives. Such arguments might be more persuasive if the present standards system served more than a symbolic function. Moreover, to the extent that OSHA regulations are effective, they are not in society's overall interest since they subvert market forces by mandating risk levels irrespective of the cost involved. OSHA's antagonistic approach to market forces is certainly misguided since the incentives for health and safety provided by the market are several thousand times larger than those created by OSHA's enforcement effort. Policies designed to augment market forces rather than to supplant them should enhance worker welfare more efficiently than will those wedded to the illusion of a non-risk society.

Toward Improved Regulatory Policies

Although the details of the needed reforms in risk regulation policies vary across agencies, these reforms share many common elements, most notably the need for policies to reflect the risk preferences of those protected and the costs imposed on individuals and firms. The pivotal institutional framework for bringing about these changes is the regulatory oversight process. The White House review process could potentially serve as the mechanism for redirecting risk regulation efforts and for providing the impetus for legislative reforms. Compared to the scope of the reforms I have suggested for OSHA, my proposals for modifying the regulatory oversight process will appear relatively modest. However, these modifications should be coupled with a revamping of the legislative mandates of risk regulation agencies. This interrelated set of proposals will constitute a dramatic shift in the approach to risk regulation.

The weaknesses of previous oversight approaches and the regulatory budget concept suggest that at best oversight can serve as a catalyst for regulatory reform. The regulatory agencies themselves must also shift their focus quite dramatically, with the oversight mechanism guiding these initiatives, providing an impartial review of the merits of proposed regulations, and ensuring that the mix of regulatory activities across agencies is efficient.

One shortcoming of the present effort has arisen only recently. Before the shift of regulatory oversight to OMB, CWPS had the legal authority to make public filings. By sacrificing this authority, the group has lost its ability to make its views public in the rulemaking process. This change limits any efforts to foster public debate, to make known the reasons for regulatory policies, and to establish the legitimacy of the oversight process. Moreover, publication of the staff's findings provided an incentive to impose higher standards on staff critiques of regulations than otherwise would be the case. Congress should give the present oversight group the authority to keep the public apprised of the basis for regulatory actions.

All previous oversight efforts have shared certain limitations. Although cost data are believed to be relatively unbiased, at present there is no *ex post* evaluation to ascertain whether these estimates have any systematic deficiencies. For example, do new regulations lead to innovative responses that reduce cost levels below the levels anticipated? Or conversely, do anticipated cost reductions fail to be realized? The data on policy benefits are even less adequate. In some cases, even the estimates of the physical effects of regulations are sketchy, and only in very unusual instances are health and safety benefits converted into monetary equivalents.

More generally, the impacts of regulatory policies should be evaluated more systematically. After-the-fact evaluation has been all but absent, as the emphasis on new regulatory proposals has imparted a prospective bias to the entire oversight process. Though there are a couple of notable exceptions, most analyses of OSHA's impact have been performed by independent academics. Other risk-related agencies have been even less thoroughly scrutinized. No White House oversight group has made a systematic effort to undertake policy evaluations of this type.[2] A potentially productive expansion of the regulatory oversight role would be to establish a policy evaluation staff that would undertake these *ex post* assessments and would provide support to analysts outside the government.

The evaluation function might be profitably coupled with an expanded scope of regulatory oversight. The present focus of these reviews is primarily on new proposed rulemakings, some of which may be sunset actions planned by the agency. The failure to delve into the existing body of regulation substantially limits the potential impact of the oversight process. Although OSHA was almost completely dormant in terms of new regulations from 1979 to 1982, it continued to enforce a large existing body of regulations, so the emphasis on oversight only for new proposals allowed the agency to remain almost immune to White House scrutiny.

A regulatory budget by its very nature establishes a constraint on all regulatory activities, so a comprehensive review of existing regulations is intrinsic to the approach. Even in the absence of a comprehensive budgetary approach, there should be increased emphasis on reassessing existing regulations. After a regulation is adopted, there is usually additional information about its effectiveness as well as the possibility of new supporting scientific evidence. Changes in related government policies may also raise coordination issues. In the absence of any external impetus for comprehensive review, there has tended to be little regulatory reform except in extreme instances, such as the elimination of the "nitpicking" OSHA safety standards.

A useful adjunct to a strengthened and expanded oversight effort would be the development of independent sources of scientific data on health risks. At present, the regulatory agencies usually summarize and interpret the health implications of the available scientific evidence on carcinogen exposure levels. The participants in the review process seldom have the capacity to assess the veracity of the regulatory agency's claims. Though these claims are sometimes questioned, the current oversight group does not have access to independent scientific evidence on these issues. The establishment of a scientific support unit, preferably independent of the Executive Office of the President, could potentially provide more balanced judgments on these issues.

A final consideration is the legislative mandate of the regulatory agencies. OSHA, for example, is legally obligated to set standards that will "assure safe and healthful working conditions." While it has long been hoped that the courts would interpret OSHA's responsibility in a more balanced way to include benefit-cost tradeoffs, the cotton dust decision points in the opposite direction. This disappointing outcome should serve as a general signal to those who were hoping that the judiciary would reform regulation after Congress and the executive branch had failed to do so. In particular, the Supreme Court thus far has not imposed any benefit-cost tests on risk regulation agencies in instances where their legislation did not.

Unless the risk regulation agencies' enabling legislation is modified it is doubtful that an executive order or a regulatory reform act requiring a budget or benefit-cost assessment can be fully effective. An agency such as OSHA could correctly argue that it is obligated to pursue quite different policies than those permitted by the oversight group. In the case of a regulatory budget, the agency might issue the regulations that it believes are needed to fulfill its mandate irrespective of any attempt to restrain its efforts. This problem is common to all risk-regulation efforts and may be even more severe in other instances, notably the Delaney Amendment's absolute requirements for FDA carcinogen regulation and the Clean Air Act's exclusion of cost considerations for the EPA ambient air quality standards.

A final means for promoting efficient regulations is through the appointment of better administrators; however, these officials will remain circumscribed by the agency's legislative mandate, its historical functions, and its constituency. But even in the wake of the recent cotton dust and benzene decisions of the Supreme Court, an innovative administrator may have considerable maneuverability. OSHA safety standards, which warrant the most extensive revision, do not pertain to toxic substances or harmful physical agents and, as a result, can be subjected to a benefit-cost test in a thorough sunset review. Health standards also can be designed in a more efficient manner through cost-effectiveness tests, stringent tests for the significance of the risks, and selection of the targets for regulation according to their likely benefits and costs. Finally, there are few legislative limitations on OSHA's ability to alter its enforcement mechanism.

I have coupled all of these policy prescriptions with caveats regarding their potential effectiveness. These reservations are based largely on the disappointing results of present risk regulation policies. There are two principal reasons why our experience with risk regulation and with regulatory reform has fallen so short of any reasonable standard of performance. First, the basic approach has been highly simplistic, with little regard for balanced consideration of the economic merits of alternatives. After policymakers identified the existence of market inadequacies in the handling of risks, which created little controversy among economists, Congress established agencies such as OSHA with quite specific mandates, without considering the other impacts of these policies. The agencies' response has been almost equally narrow, issuing stringent standards in an attempt to abolish hazards. To the extent that economic considerations entered at all it was usually in terms of technical feasibility. The enforcement of the standards, which provides few financial incentives for safety but nevertheless serves as a form of systematic harassment of the business community, also has suffered from a failure to incorporate economic considerations into policy design. The fundamental questions that should have been addressed are how will the regulatory policy affect market behavior and, given this response, how should the policy be structured?

The second hindrance to effective policies is more fundamental and is inherent in all policy options. The chief reason why the market does not handle risks adequately is that participants are not fully informed of the risks associated with their choices. In some cases the government may have different or superior information about a potential hazard, thus giving it some potential advantage, but usually substantial uncertainty remains. Carcinogenic risks in particular are not sufficiently understood to enable us to enact policies to achieve efficient market outcomes. The most that can be hoped for is that policies will be based on the best information currently available and that these policies will be revised as we learn more about the implications of different risks and the market responses to them.

GENERAL RAMIFICATIONS

The principal implications of this analysis of job risks are pertinent to many other forms of risk associated with market processes. While the evidence suggests that diverse market mechanisms promote the handling of risks in a systematic fashion, there are obvious impediments to the efficacy of the market. Chief among these shortcomings is imperfect information regarding risks. Better information will lead workers to quit or consumers to switch products as they learn about the risks involved. Even with these additional market responses, there may be inadequacies, leading to a bias toward overly risky activities and dimly understood technologies in a wide variety of instances. Few observers would claim that government regulation has no potentially productive role to play in these situations. Unfortunately, this potential has not been realized because of the fundamentally misguided nature of risk regulation policies. Indeed, in its decade of existence, OSHA has established a reputation as perhaps the most ill-conceived governmental intrusion in the marketplace.

If risk regulation policies are ever to realize their potential, they must be based on an understanding of the market effects of different risks and the effects of regulatory policies on choices by individuals and firms. Well-designed policies should not eliminate the diversity of risks and strive for a no-risk society, but instead should provide risk information and promote the economic incentives that will lead to the wage-risk and price-risk combinations most appropriate for different risk preferences. Any policy addressing market-traded risks should rely on individual choice. The substantial heterogeneity in people's attitudes toward risk suggests that there will be serious welfare losses if the range of options is greatly reduced. Moreover, the incentives for safety created by the expression of these choices in the marketplace are much more powerful than even a greatly expanded enforcement of present risk standards.

Proposals to rationalize the regulatory process in this manner may be attacked as irresponsible and immoral since, in effect, lives will be traded off explicitly for dollars. But tradeoffs of this type are made implicitly by individuals and by all government policies that could have produced additional health benefits if more resources had been expended. The most compelling point is that the absolutist approach to regulation, with rigid standards intended to eliminate risks, has yet to produce significant improvements in workers' well-being but has nevertheless inflicted substantial burdens on firms. Perhaps the greatest losses from the present approach are the deleterious health effects on workers that could have been prevented if risk regulation policies had not been conceived as independent of the market context in which they operate.

Efficient regulatory policies promise to be more beneficial to society as a whole and to those exposed to risks since the degree of regulation will reflect the preferences of those whose lives are being protected. The justification for the current absolutist policies rests on the illusion that society should not and, at present, does not trade off dollars for lives. The fundamental question that we must address is whether we should continue to sacrifice additional lives in an effort to preserve this popular mythology.

Appendix

Notes

References

Index

APPENDIX

ECONOMIC IMPLICATIONS OF
THE QUIT PROCESS

THE STRENGTH OF THE QUIT RESPONSE BY WORKERS WHEN THEY learn about job risks indicates that even with imperfect information market mechanisms do function systematically. To investigate whether this behavior eliminates inadequacies in the market, we need to delve into the conceptual underpinnings of the model underlying my discussion in Chapter 4.

The focus of this appendix is on workers on a continuing job in which they face a sequence of risks over time. A principal abstraction involved in the analysis is the assumption that anticipation of future risks does not reduce workers' welfare by increasing their anxiety. The pivotal role of information in generating anxiety and suspense in cinematic situations has been emphasized by Alfred Hitchcock: "In the usual form of suspense it is indispensable that the public be made perfectly aware of all of the facts involved. Otherwise, there is no suspense."[1] Although a similar response may diminish the value of risk information to workers, I will not consider this here.[2]

Since a worker's behavior may run counter to what we would believe intuitively, it is helpful to delve a bit into the economic underpinnings of his decisions. The worker's choice is akin to the classic two-armed bandit problem in which a gambler is engaged in a series of trials with a choice of two slot machines.[3] On each trial the chances for success are known for one machine and are uncertain for the other. The gambler's problem is to decide whether he wants to experiment with the uncertain machine, and if so, what its minimal acceptable performance must be after different numbers of trials for him to wish to continue experimenting. A worker considering a job with uncertain properties faces a similar choice. He must decide whether to start work on that job and then, as he learns about its characteristics, he must decide whether he will quit and switch to a different position.

In the discussion below, I will draw a distinction between the level and the precision of workers' risk perceptions.[4] The level of the risk perception pertains to the assessed probability of the outcome on the next trial, which in the case of employment is the next work period. The precision of the worker's probabilistic beliefs pertains to the level of information he believes he has about the risk. Workers with tight probability assessments act as if they are basing their judgments on a long series of experiments. Additional information acquired on the job will alter tight priors very little since these workers place substantial confidence in their initial beliefs. The perfect information situation can be viewed as the limiting case of precise risk perceptions. The value of learning on the job is greatest for workers whose risk perceptions are loosest.

The role of the precision of workers' probabilistic beliefs is illustrated by the following situation. Suppose job 1 offers a 0.5 probability of success (remaining uninjured) in each period and an equal probability of failure (being injured). These probabilities are precisely known. Experimentation with that job consequently will not affect the worker's future assessment of the risk it entails. The properties of job 2 are uncertain. Although the worker believes it offers an initial 0.5 probability of success, he will revise this probability upward after each success and downward after each failure as he learns by experience about the properties of the job. Suppose there is more than one period to the worker's choice problem and that he observes either a success or a failure in each period for the job he selects. Finally, a success on either job yields an identical wage, while a failure (injury) leads to a less preferred outcome that is equally unattractive for either job. Which job should the worker select, job 1 with the "hard" 0.5 probability of success, or job 2 with the "soft" 0.5 probability based on the worker's subjective assessment?

Table A.1 summarizes the worker's probabilistic beliefs. For job 1, the chance of remaining uninjured is 0.5 in each period, irrespective of the job outcome. For job 2 it is increased after a first-period success to some value, say 0.67, while an injury in period 1 leads the worker to revise the probability downward to 0.33.[5] The optimal employment decision is to start work on the uncertain job, continue on it if the outcome in period 1 is favorable, and switch to job 1 if the outcome is unfavorable.

These properties are quite general. If there is a sequence of trials, it is always desirable to start with the uncertain job so long as the initial chance of success is not too far below the corresponding value for job 1. The worker in effect displays a systematic preference for jobs whose properties are not fully understood. The reason for this preference is the potential for learning and either quitting if the outcomes are not sufficiently favorable or remaining with the job if he learns that the risks are low.

So long as the outcomes on the uncertain job are successful, the worker will not leave it; he will leave only if the outcomes are sufficiently adverse. Although it is always desirable for the worker to "stay on a winner," how soon he chooses to leave a job that appears to be a loser depends on a variety of factors, especially how quickly he revises downward the chance of remaining healthy on that job.

The worker who quits his job may not necessarily think that the job is now more unattractive than he expected initially. He might believe there is an equal chance that the job is very safe, very risky, or of average riskiness. Even if the worker's initial

TABLE A.1. DATA FOR THE JOB RISK CHOICE PROBLEM

Job	Probability of injury	
	Period 1	Period 2
1	0.50	0.50
2	0.50	0.67 if injured in period 1
		0.33 if not injured in period 1

expectation is that the position poses average risk, he may choose to stay only if he learns that the position is very safe. This phenomenon occurs in other situations as well. Aspiring actors may not expect to be major stars, but they may be eager to try this profession for a short period in order to gain information about what they initially view as a small probability of a large reward.

An intrinsic characteristic of this on-the-job experimentation is that a worker will only leave a position whose risks are known with precision if some other aspects of his opportunities have changed. Since the worker initially has all risk information about such jobs, he would have no reason to quit once he started work on it.

The substantial empirical support for the job hazard–quit relationship consequently is very strong evidence for the view that workers begin jobs with imperfect information, learn about the risks on the job, and then quit if they acquire very unfavorable information. After controlling for other factors, such as workers' health status, we find that it is this learning effect that contributes to the independent relationship between quitting and the risks of employment. The substantial magnitude of the effect derives in part from the fact that workers will display a systematic preference for jobs whose implications are not fully understood; that is, a risk that is not known with precision will be preferred to equal or slightly larger assessed risks that are known precisely.

The learning process described thus far allows for the possibility of a nonfatal job injury. However, the event that conveys job risk information may be a permanently maiming injury or an irreversible ailment. The extreme example is the risk of death. Suppose job 1 offers a 0.5 probability of being killed and an equal chance of staying alive during each of the two periods. As before, this probability is known with precision. Job 2 has uncertain properties, but the worker assesses his chance of being killed in the initial period as being identical to that of job 1. The potential for quitting, which in the earlier example led the worker to prefer the job with uncertain properties, is clearly irrelevant since the worker is dead after any adverse job outcome and would not

TABLE A.2. DATA FOR THE DEATH RISK CHOICE PROBLEM

| Job | Probability of not being killed | | |
	Period 1	Period 2	Survival for two periods
1	0.50	0.50	0.25
2	0.50	0.67	0.33

leave a job he started after a successful outcome (staying alive). If the jobs are otherwise identical, should the worker pick the job posing known risks, select the uncertain job offering the same initial risk, or view the jobs as being equally unattractive?

In this case, as earlier, he prefers the uncertain job. Table A.2 summarizes the information most pertinent to the worker's choice. Each job offers the same chance for survival in the initial period. The chance of not being killed on job 1 in either of the two periods simply compounds this risk, yielding an overall chance of survival for two periods of 0.25. On job 2, however, if he is not killed initially, the worker learns that the job is less risky, thus raising his expected chance of remaining alive in period 2. His chance of surviving on the job for both periods is 0.33, which is better than his prospects on job 1. The only pertinent information the worker receives on job 2 is good information. Adverse information acquired by suffering an irreversible event, such as death, does not enter into consideration since it does not affect any future employment decisions.

When a person faces a sequence of lotteries with irreversible consequences, loose prior beliefs about the risk tend to be preferable since lotteries with imprecisely known risks offer the greatest chance for long-term survival.[6] This predilection for jobs whose implications are not fully understood has several fundamental implications for the role of governmental intervention, which are described in Chapter 5.

NOTES

2. OSHA: DESIGN, IMPLEMENTATION, AND IMPACT

1. Section 26 of the Occupational Safety and Health Act of 1970, 29 U.S.C. § 651 (1976), hereafter referred to as OSHA Act.
2. OSHA Act, section 2b, part 3.
3. OSHA Act, section 5a, part 1.
4. OSHA Act, section 3b, part 7; section 6b, part 5.
5. See the decision of the Supreme Court in *Industrial Union Department, AFL-CIO v. American Petroleum Institute,* 448 U.S. 607 (1980), hereafter referred to in the text as the OSHA benzene decision, and the Supreme Court decision in *American Textile Manufactures Institute v. Donovan,* 452 U.S. 490 (1981), hereafter referred to as the cotton dust decision.
6. See the OSHA benzene decision.
7. See MacAvoy (1979) for a detailed presentation of this view and a historical perspective on other regulatory agencies.
8. The actual number of standards varies with how one chooses to count the various subparts of each standard. The general industry standards were published in *Code of Federal Regulations,* part 1910, title 29. The current version of these standards runs to 820 pages even after the elimination of the "nitpicking" standards. See U.S. Department of Labor, Occupational Safety and Health Administration (1979b). (OSHA publications are listed in the References under U.S. Department of Labor; hereafter in the notes they will be referred to as OSHA with the date.)
9. This anecdote has been popularized by Paul MacAvoy, who was a member of the President's Council of Economic Advisers and the chief regulatory official in the Executive Office of the President at the time of this incident.
10. This justification is stated quite explicitly in the briefing notes prepared for all OSHA officials. See OSHA (1980b), Standards Deletion section.
11. See OSHA safety and health standards, *Code of Federal Regulations,* part 1910, title 29, subpart Z.
12. There have been a number of analyses of OSHA, some of which address such issues more fully. These studies include: Ashford (1976), Bacow (1980), Mendeloff (1979), Oi (1975), Page and O'Brien (1973),

R. Smith (1976), Zeckhauser and Nichols (1979), and Morrall (forthcoming).

13. These personnel figures understate the number of employees administering the program since, traditionally, roughly half of all states have operated their own OSHA-approved inspection plans. In fiscal year 1981, OSHA's share in the cost of these plans was $42 million.

14. These priorities are specified in OSHA (1979a), p. II–1.

15. For the earlier priorities, see OSHA (1973), *The President's Report on Occupational Safety and Health,* pp. 3–4.

16. The figures discussed below are based on unpublished OSHA computer printouts supplied to the author.

17. These patterns are distorted somewhat by the shift in the federal fiscal year in FY 1977.

18. See OSHA (1980b), pp. I–10 and I–11.

19. See OSHA (1979a), p. VIII–1.

20. See OSHA (1979a), pp. VIII–5 and X–3.

21. See OSHA (1979a), pp. VIII–9 and VIII–10.

22. These guidelines are specified in OSHA (1979a), pp. XI–1 and XI–2.

23. These calculations are based on data from OSHA computer printouts, some of which appear in Table 2.3.

24. See, for example, OSHA (1973) *President's Report,* p. 1.

25. These case hour figures are based on OSHA computer printouts for FY 1980.

26. These criticisms are documented in Oi (1975). An update on these issues using Wisconsin data appear in U.S. Department of Labor, Office of the Assistant Secretary for Policy, Evaluation, and Research (1980). The data in this and the following paragraph are based on this report.

27. This section provides a nontechnical discussion of the model presented in Viscusi (1979b).

28. As I indicate in Viscusi (1979b), this aberrational result will not occur if firms come into compliance with the standard before this counterproductive region is reached.

29. The Wisconsin data are from Oi (1975); the 1931 data are from National Commission on State Workmen's Compensation Laws (1973), pp. 287–288; and the North Sea data are from U.S. Council on Wage and Price Stability (1977) filing on the OSHA diving standard. For an interesting discussion of the implications of these earlier studies for the limits on the potential risk reduction achievable under OSHA, see Zeckhauser and Nichols (1979).

30. These studies are summarized in National Commisssion on State Workmen's Compensation Laws (1973), pp. 287–288.

31. See OSHA (1980d) on fatalities on oil and gas well drilling rigs.

32. Although one can calculate marginal products and average products of each input, such calculations are much more sophisticated than the assignments of responsibility discussed above.

33. See Page and O'Brien (1973) for a description of such efforts to place the responsibility for most risks on workers.

34. See the excerpts from the House and Senate debates in Bureau of National Affairs (1971).

35. The BLS sample before 1971 was based on voluntary industry reports, with about 49 percent of all manufacturing firms responding in 1970; the sample of reporting firms changed annually, potentially distorting risk trends. The injury rate measure used is the number of disabling injuries per million hours worked. To be disabling, an injury must have caused permanent damage or the loss of the ability to work at a regularly established job for a complete day following an injury.

36. The NSC data are based on voluntary reports by firms affiliated with NSC.

37. The new injury rate measure was the number of recordable injuries per hundred full-time employees. Recordable injuries include fatalities, lost workday cases, and nonfatal cases without lost workdays.

38. To reduce the paperwork burden on small firms, the BLS abolished injury reporting requirements for enterprises with fewer than eleven employees in 1978. As will be noted below, however, the BLS death rate series from 1973 to 1979 is calculated for a consistent sample of firms with eleven or more employees. The NSC measure also changed, because of a modification in the manner in which it calculated the number of exposed workers.

39. The prominence of this alleged increase is discussed in Page and O'Brien (1973) and R. Smith (1976), among others. Ashford (1976) discusses these trends and places great emphasis on their implications for policy. Oi (1975) presents a thoughtful discussion of whether it was the injury rate level or its trend which led to OSHA.

40. A limitation of these data is that it is somewhat arbitrary to classify an investment as being related to health and safety. Since this variable serves as a dependent variable, random measurement error will not bias the results discussed here.

41. The patterns in the update of Mendeloff's work are based on a discussion I had with him in 1980 regarding his more recent work.

42. In particular, the gap between actual and predicted injury rates

after 1970, which was obtained using a pre-1971 equation, in no way implies that the gap was due to OSHA. One might also raise a variety of statistical issues concerning the validity of the results, irrespective of their interpretation: 1) the lagged injury rate value is included in the injury rate equation, and if the error terms are serially correlated, this lagged value and the contemporaneous error term will be correlated; 2) a time trend variable was included, but nonlinear time trends were not, leading to possible misestimation of the temporal effect for purposes of postsample period simulation; 3) accession rates and new hire rates were included as explanatory variables even though the causality may be in the other direction due to the job hazard–quit relationship; and 4) California dependent variables (injury rates) were coupled with U.S. explanatory variables. The existence of these problems does not imply that Mendeloff's results should be dismissed, but it does suggest that one should not place too much weight on even the few instances in which he finds patterns favorable to OSHA.

43. In an earlier analysis, Smith (1976) found no evidence of any impact of OSHA's Target Industry Program—a small-scale effort to direct inspections at a selected group of industries.

44. This relationship in turn has produced a small significant link overall for the entire sample in 1973.

45. In my own model (Viscusi, 1979b) this variable serves as a proxy for both the stock of health and safety capital and the optimal riskiness of the technology for the firm. Although the net impact of these conflicting influences is not clear *a priori*, on an empirical basis the lagged injury rate bears a strong positive relationship to the current risk level.

3. Compensating Differentials for Risk

1. Although some people may find it enticing to encounter risks, this behavior is not the norm. Nevertheless, the compensating differential theory can be generalized to include such complications. See Viscusi (1979a), chap. 5.

2. The discussion in this chapter is based principally on Viscusi (1979a), chaps. 2 and 15. Other contributions to this literature include Oi (1973, 1974), Thaler and Rosen (1976), and R. Smith (1976). These works have created a modern revival of interest in Adam Smith's classic analysis of compensating differentials.

3. The job risk figures are based on the BLS data discussed below, while the risks for nonjob exposures are based on the calculations of

Harvard physicist Richard Wilson in "A Rational Approach to Reducing Cancer Risk," *New York Times,* July 9, 1978, p. E17.

4. Supporting data on noise exposures can be found in *Design News,* March 3, 1975, pp. 52–55.

5. Smith (1776; reprinted. 1937), pp. 99–100.

6. See, especially, pp. 87, 116, 159, 187, and 288 of Engels (1845; reprint ed. 1968).

7. The impact of insurance market inadequacies on the risk level is often quite complex since it hinges on the nature of the inadequacy, the accuracy of workers' risk perceptions, and whether or not the job injury loss is purely financial. In some instances, poorly functioning insurance markets could lead to an inefficiently low level of risk since firms must compensate workers through wage premiums for their inability to insure.

8. Namath, interview with Johnny Carson on "The Tonight Show," July 20, 1979.

9. *New York Times,* July 3, 1979, p. A6.

10. I examined the full-time blue-collar workers in the University of Michigan Survey of Working Conditions. These results are based on Viscusi (1979a).

11. These other studies do not, however, explicitly link risk premiums to workers' perceptions.

12. See Viscusi (1979a), chap. 15. The $900 premium is based on the coefficient of the hazard premium variable in an earnings regression equation.

13. This estimate is based on 1979 injury rate data using a $2 million value of life and a $20,000 value of injuries. These amounts are multiplied by the number of cases of each type to obtain the total level of risk premiums in the private sector. See Chapter 6 for supporting data on the implicit values.

14. This compensation payment estimate is also for 1979. See National Safety Council (1981), p. 39.

15. My categorization procedure for health risks and safety risks is discussed in Chapter 4.

16. This relationship is derived formally in Viscusi (1979a), chap. 2.

17. The supporting statistical analysis is presented in Viscusi (1979a), chap. 14.

18. Average work hours data are from U.S. Bureau of Labor Statistics (1981), *1980 Handbook of Labor Statistics.*

19. These accident data are summarized in National Safety Council (1979).

20. An interesting long-term perspective on the evolution of U.S. regulation is provided by MacAvoy (1979).

21. See Ashford (1976), chap. 11, especially p. 517, for advocacy of international standards for risk and import bans for riskily produced goods.

22. For fuller treatment of these issues, see Viscusi (1979a), chaps. 11, 14, and 15.

23. For union effects on wage premiums see Thaler and Rosen (1976) and Viscusi (1979a, 1980d). Union influence on fringe benefit coverage is discussed in Goldstein and Pauly (1976), Freeman (1981) and Viscusi (1979a), chap. 15.

24. This estimate is based on Viscusi (1979a), chap. 15.

25. See Viscusi (1979a), chaps. 14 and 16, for supporting evidence.

4. Market Forces and Inadequate Risk Information

1. These results are based on the analysis in Viscusi (1979a), chaps. 14 and 16, using the University of Michigan Survey of Working Conditions. The sample includes 496 full-time blue-collar workers.

2. The average industry risk figures were calculated using the pre-OSHA risk figures gathered by the U.S. Department of Labor's Bureau of Labor Statistics, and these were matched to workers in each subjective risk perception group. The risk is in terms of disabling injuries per million hours worked.

3. The small sample size and differences in jobs within industries no doubt contribute to the slight departure from a pattern of steady increase.

4. The model of worker quit behavior induced by learning about job risks was introduced in Viscusi (1979a), chap. 4. Chapter 13 of that volume presents the first empirical investigation of the impact of nonpecuniary job characteristics on worker quitting. All information in the rest of this chapter is drawn from these chapters in my earlier book, except where indicated otherwise.

5. Changes in the worker's health status due to job injuries may, of course, alter the attractiveness of the job. The empirical results for individual quit behavior control for worker health status.

6. Underlying my discussion is an assumption that individuals assess subjective probabilities and modify them in Bayesian fashion based on their experience. Raiffa (1968) provides an eloquent defense of the Bayesian decision theory framework from a normative perspec-

tive. Whether individuals actually behave in a manner that is consistent with this model is a quite different empirical issue, which is addressed in this chapter and in Viscusi (1979a).

7. Although worker injuries are associated with hazard perceptions even after taking into account an extensive variety of job characteristics, only an empirical test that explicitly assesses the actual changes in workers' risk perceptions can be conclusive. Workers on hazardous jobs are more likely to perceive their jobs as risky and are more likely to be injured. When we focus on the perceptions of injured workers, we may not be capturing a learning effect but instead may simply be identifying workers on risky jobs more accurately.

8. This analysis utilized the University of Michigan's Panel Study of Income Dynamics.

9. See Viscusi (1980b) for an analysis of the limitations of the wage structure as a self-selection device in this situation.

10. For a formal analysis of the decision facing the employer and supporting empirical evidence, see Viscusi (1979a), chaps. 7, 8, and 14.

11. See Viscusi (1980a) for a derivation of this result.

12. The influences discussed below are analyzed more rigorously in Viscusi (1979a), chaps. 8 and 9.

13. An illuminating discussion of this and other deviant properties of information as an economic good can be found in Arrow (1971).

14. In Viscusi (1980a) I derive each of these results.

5. THE BASIS FOR GOVERNMENT INTERVENTION

1. For a lucid discussion of these issues see Alpert and Raiffa (1969). A more general discussion of probability assessment appears in Raiffa (1968).

2. The problem of pyramiding intervention and the importance of a variety of other externalities is discussed in much more detail by Zeckhauser and Nichols (1979), who place greater emphasis on the externality-oriented basis for intervention than I do in this study. A survey of the limited evidence on the magnitude of these externalities is provided by Morrall (forthcoming).

3. Kelman (1981) is a principal advocate of this view.

4. Many of these forms of market failure are analyzed in Viscusi (1979a). Two notable problems are that employers may have difficulty in monitoring the riskiness of different workers and their actions, making job assignments difficult, and workers may have limited job mobil-

ity, which will impede their attempt to leave a hazardous job.

5. More specifically, below the level where marginal benefits equal marginal costs, the incremental benefits from increasing the scale or stringency of the policy will outweigh the incremental costs, implying that greater net benefits (benefits less costs) can be produced by increasing the intensity of the program. Similarly, above the optimal level, marginal costs exceed the marginal benefits, implying that a reduction in the policy effort will yield greater net benefits. For further discussion of these issues, see a public policy analysis text such as Stokey and Zeckhauser (1978).

6. The persistence of these biases is legendary. In an earlier study, Richard Berkman and I document the extent of these distortions, which is considerable. See Berkman and Viscusi (1973).

7. This property is usually referred to as the "potential Pareto compensation principle."

8. This notion is the central theme of Lester Thurow's (1980) analysis. Although Thurow discussed regulatory policies briefly, the misguided notions discussed below are those of individuals who have extrapolated Thurow's views on aggregate economic policies to the quite different case of risk regulation.

9. The analytics of this unraveling process are detailed in Viscusi (1979a).

10. For a more technical discussion of the issues raised in this section see Viscusi (1980a, 1979a). Recent comprehensive studies of the workers' compensation system include the analyses by Chelius (1977) and by Darling-Hammond and Kniesner (1980).

11. This assumption is incorporated in my theoretical work as well as in various medical insurance analyses. Although the predictions of models using this assumption have been borne out, as yet there has been no refined empirical test of the claim that injuries and illnesses reduce the marginal utility of income.

12. See the summary of these plans in U.S. Chamber of Commerce (1980).

13. See National Commission on Workmen's Compensation Laws (1973), pp. 36–37.

14. A final overriding feature of workers' compensation is that these programs are administered by the states. This decentralization could potentially promote increased policy innovations and greater responsiveness to the preferences of citizens in different areas. There is also a major disadvantage in that there has been very little coordination of

state-run policies with federal risk regulation efforts, and there is no pooling across states of the valuable information generated by a workers' compensation program. If there were greater centralization of the risk and insurance data, policy analysts could potentially identify particularly effective policies and shifts in risk patterns. At present, very few states have made this information available, which may have contributed to the comparative lack of attention given to these programs.

6. THE VALUE OF LIFE AND LIMB

1. This approach was introduced in the classic essay by Schelling (1968).

2. See, for example, Rice and Cooper (1967) for an early example of the income-based approach.

3. See Acton (1973) and Jones-Lee (1976) for examples of the interview method.

4. For a more comprehensive survey of these studies, see Smith (1979). In Viscusi (1978) I present a critical review of the major conceptual and empirical studies in this area. Related work in this area includes the survey by Bailey (1980).

5. The interview study by Jones-Lee (1976) also had a relatively small sample size—thirty. The Jones-Lee study raises an additional self-selection problem since only thirty of the ninety people polled responded.

6. A major difference in the equations estimated is the nonpecuniary rewards variables included. Thaler and Rosen's analysis and that of Smith focused on death risks alone. (Smith's estimate of premiums for nonfatal injuries were insignificant.) In both of my analyses, the wage premiums for fatal risks were distinguished from those for nonfatal risks. Moreover, in my analysis of the Survey of Working Conditions data I took into account a diverse group of other nonpecuniary characteristics of the job as well as whether or not workers perceived any risk. The inclusion of these variables should increase the degree to which we are estimating wage premiums for risks rather than rewards for other unpleasant job characteristics correlated with riskiness.

7. In particular, the risk variable assumed a value of zero if the worker did not view his job as hazardous in any respect.

8. Because differences among jobs within an industry are ignored, these variables are subject to undoubtedly large measurement prob-

lems, leading to an underestimate of risk premiums if this error is random.

9. The risk measures were based on the BLS death risk for the worker's industry and the nonfatal lost-workday injury and illness rate for the industry. These risk variables differ from those used in my earlier study since the earlier variables were reported to BLS on a voluntary basis, while the more recent variables are reported on a mandatory basis by all firms. In addition, there were some changes in the injury rate definitions. Despite these differences, the annual death risk and nonfatal injury risk were roughly identical for both of the samples that I analyzed.

10. In particular, I estimated the effect of death risks and the square of death risks on worker wages. No significant nonlinearities were present for nonfatal injuries.

11. Equation 1 used the worker's wage rate as the explanatory variable, while equation 2 used the logarithm of the wage rate.

12. In addition to *ex ante* wage compensation, workers also received substantial tax-free workers' compensation benefits—up to two-thirds of the worker's gross wages in most states—so that the injury results in particular underestimated the implicit values attached by workers.

13. Regulation of job hazards also imposes implicit costs on employers. The amounts of these costs are often difficult to compute directly, since firms have an incentive to overstate the financial burdens of prospective regulations. On a theoretical basis, however, the marginal cost of safety improvements in a competitive market equals the cost of the wage premiums for an incremental change in the level of the risk. The amount by which wages are reduced and costs increased depends on a variety of factors such as the level of risk and the extent to which it is reduced, the types of hazard, and characteristics of the workers and the workplace, such as unionization. Nevertheless, applying the implicit values of life and limb to analyses of the impacts of health and safety regulation might be a useful starting point for policy evaluation.

14. For a detailed discussion of measures of quality-adjusted lives, see Zeckhauser and Shepherd (1976).

15. Two important exceptions should be noted. If beneficiaries systematically misallocate resources and neglect the future, a case might be made for using implicit values of life above those revealed in the marketplace. Moreover, if there are substantial externalities, the preferences of society as a whole must also be incorporated in the analysis.

16. In particular, see Mishan (1971).

7. How to Set Standards If You Must

1. The material in this section is based on Viscusi (1979b) and Viscusi and Zeckhauser (1979).

2. The OSHA carcinogen policy was issued in *Federal Register*, vol. 45, no. 5, book 2, Jan. 22, 1980, pp. 5001–5296. For a detailed comparison of the OSHA and EPA carcinogen policies, see the analysis by Broder and Morrall (1980).

3. These estimates were converted into current prices using cost data presented in U.S. Council on Wage and Price Stability (1978b).

4. See *Federal Register*, vol. 45, no. 157, Aug. 12, 1980, pp. 53672–53679.

5. Two such models are the logit and log-probit models.

6. U.S. Council on Wage and Price Stability (1977b).

7. Although the salary levels are broadly consistent with compensating differentials, ideally one would like to undertake the kind of detailed statistical analysis that was the basis of the results in the value-of-life discussion.

8. Alternative cost estimates are presented for cotton dust in Morrall (1979) and for the four carcinogens in Broder and Morrall (1980).

9. The discussion of the arsenic standard is based on my calculations and material in Miller (1976) and Levine (1976). Using different risk assumptions, Broder and Morrall (1980) estimate much higher costs for this standard. All cost estimates for this standard are partial estimates.

10. Technical information for my discussion is based in part on information in *Federal Register*, vol. 43, no. 122 (1978), pp. 27351–27399, and on material in Morrall (1976b) and U.S. Council on Wage and Price Stability (1977a). The cotton dust standard is also discussed in Morrall (1979). Alternative analyses of the noise standard can be found in Morrall (1976b) and Smith (1976). Kryter (1970) presents an excellent discussion of the health effects of noise.

11. Since these data pertain to average rather than marginal costs, they will be used to highlight the variations in cost-effectiveness rather than to analyze where a standard in a particular industry should be set.

12. A very detailed report on the advantages of performance standards was edited by MacAvoy (1977).

13. See OSHA, (1979b), pp. 20–21, or *Code of Federal Regulations*, part 1910, title 29, 23.

14. See OSHA (1979b) pp. 421–422, or *Code of Federal Regulations*, part 1910, title 29, 244.

15. The preface of MacAvoy (1977) is the source of these estimates.

16. There may, however, be other uncertainties with engineering standards. The immense volume of OSHA standards made enterprises uncertain about which standards the agency would choose to enforce. The enforcement process has been fairly uneven, with readily monitorable hazards receiving most attention.

17. These estimates were calculated using data from Morrall (1976b).

18. The self-selection of workers to jobs may not yield efficient outcomes in many of these instances. See Viscusi (1979a), especially chap. 6.

19. Some of these rigidities may reflect quite reasonable economic behavior. For small productivity differences, it may be quite costly both administratively and in terms of morale to vary the wage rate for a job.

20. Workers will, of course, receive *ex ante* wage premiums to reflect the expected number of accidents that will be inflicted upon them. However, this mechanism seldom, if ever, reflects the risk associated with one's own job. If one worker in a hundred is careless, the overall risk for the firm is small, but if he is working on a job next to you, your risk may be quite high.

21. One prominent factor influencing these costs is whether workers are being excluded from new jobs or being moved from their present jobs. The costs to the firm and the worker are typically greater in the latter instance.

8. The Political Context of Risk Regulation Policies

1. For different perspectives on the regulatory review process, see MacAvoy (1979), who focuses on the Ford administration, and DeMuth (1980a, b), who focuses on regulatory budgets as well as on the Council on Wage and Price Stability.

2. These calculations are based on data appearing in U.S. Council on Wage and Price Stability (1978b), p. 10.

3. For an alternative discussion of how the establishment of the stakes and stands of the players in the regulatory process influences policy outcomes, see DeMuth (1980a, b).

4. This action seems to have been motivated primarily by a desire to terminate the CWPS pay/price standards program rather than by an attempt to alter the regulatory oversight process. Abolition of the

agency, rather than scaling it down by 80 percent, had obviously greater political appeal as well.

5. For example, the choice of whether to establish a regulatory budget in OMB or a review unit under CEA that strengthens the RARG approach may hinge on which group exercises greater power over economic decisions. A principal consideration is whether the head of the regulatory review also chairs Cabinet-level economic policy discussions (through the Economic Policy Group under Carter or the Economic Policy Board under Ford), since the chairman of this group has greater ability to define the policy agenda and to promote his own policy interests.

9. Controlling Risks through Individual Choice

1. See OSHA (1981).

2. Some CWPS analysts, most notably John Morrall, did, however, delve more deeply into such issues. See, for example, Broder and Morrall (1980) and Morrall (1979 and forthcoming). These analyses are more in the spirit of a critique of proposed policies and a synthesis of existing research than an effort to assess new evidence regarding OSHA's impact.

Appendix. Economic Implications of the Quit Process

1. See the classic interviews with Hitchcock in Truffaut (1967).

2. A detailed economic analysis of these effects is provided in Viscusi (1979a), chap. 5.

3. For a technical treatment of two-armed bandit problems, see Berry (1972).

4. Below I will refer to priors associated with substantial prior information as "tight" priors. As Raiffa and Schlaifer (1961) emphasize, this visual simplification is not always an entirely accurate description.

5. These calculations are based on the assumption that the worker's prior probability distribution is uniform. A similar result generalizes to the entire beta family of distributions.

6. This result is analyzed in detail for the employment choice situation in Viscusi (1979a), chap. 4. Berry and Viscusi (1981) explore the theoretical properties of this generalization of the two-armed bandit framework.

REFERENCES

Acton, Jan. 1973. *Evaluating Public Programs to Save Lives: The Case of Heart Attacks.* Santa Monica: RAND Corporation.

Alpert, Marc, and Howard Raiffa. 1969. A Progress Report on the Training of Probability Assessors. Unpublished manuscript, Harvard University.

Arrow, Kenneth J. 1971. *Essays in the Theory of Risk-Bearing.* Chicago: Markham Publishers.

Ashford, Nicholas. 1976. *Crisis in the Workplace: Occupational Disease and Injury.* Cambridge, Mass.: MIT Press.

Bacow, Lawrence. 1980. *Bargaining for Job Safety and Health.* Cambridge, Mass.: MIT Press.

Bailey, Martin J. 1980. *Reducing Risks to Life: Measurement of the Benefits.* Washington: American Enterprise Institute.

Berkman, Richard, and W. Kip Viscusi. 1973. *Damming the West.* New York: Grossman Publishers.

Berry, Donald. 1972. A Bernoulli Two-Armed Bandit. *Annals of Mathematical Statistics* 43:871–897.

Berry, Donald, and W. Kip Viscusi. 1981. Bernoulli Two-Armed Bandits with Geometric Termination. *Stochastic Processes and Their Applications* 11:33–45.

Broder, Ivy, and John F. Morrall III. 1980. The Economic Basis for OSHA's and EPA's Generic Carcinogen Standard. Paper presented at 1980 Association for Public Policy and Management Research Conference.

Brodeur, Paul. 1974. *Expendable Americans.* New York: Viking Press.

Bureau of National Affairs. 1971. *The Job Safety and Health Act of 1970.* Washington: Bureau of National Affairs.

Cheit, Earl. 1961. *Injury and Recovery in the Course of Employment.* New York: John Wiley and Sons.

Chelius, James. 1977. *Workplace Safety and Health: The Role of Workers' Compensation.* Washington: American Enterprise Institute.

Darling-Hammond, Linda, and Thomas J. Kniesner. 1980. *The Law and Economics of Workers' Compensation.* Santa Monica: RAND Corporation.

DeMuth, Christopher. 1980a. Constraining Regulatory Costs—Part I: The White House Review Programs. *Regulation* 4:13–26.

———— 1980b. The Regulatory Budget. *Regulation* 4:29–43.

Dunlop, John. 1958. *Industrial Relations Systems*. Carbondale, Ill.: Southern Illinois University Press.

Engels, Friedrich. 1845. *The Condition of the Working Class in England*. Reprinted., Stanford: Stanford University Press, 1968.

Freeman, Richard. 1976. Individual Mobility and Union Voice in the Labor Market. *American Economic Review* 66:361–368.

———— 1981. The Effect of Trade Unionism on Fringe Benefits. *Industrial and Labor Relations Review* 33:489–509.

Goldstein, Gerald, and Mark Pauly. 1976. Group Health Insurance as a Local Public Good. In *The Role of Health Insurance in the Health Services Sector*, ed. R. Rosett. New York: National Bureau of Economic Research.

Gordon, Jerome, Allan Akman, and Michael Brooks. 1971. *Industrial Safety Statistics: A Re-Examination*. New York: Praeger.

Iskrant, Albert, and Paul Joliet. 1968. *Accidents and Homicide*. Cambridge, Mass.: Harvard University Press.

Jones-Lee, M. W. 1976. *The Value of Life: An Economic Analysis*. Chicago: University of Chicago Press.

Kelman, Stephen. 1981. Cost-Benefit Analysis—An Ethical Critique. *Regulation* 5:33–40.

Kinnersly, Patrick. 1973. *The Hazards of Work: How to Fight Them*. London: Pluto Press.

Kryter, Karl. 1970. *The Effects of Noise on Man*. New York: Academic Press.

Levine, Dianne. 1976. Statement on Behalf of the Council on Wage and Price Stability before OSHA, "Exposure to Inorganic Arsenic."

MacAvoy, Paul, ed. 1977. *OSHA Safety Regulation: Report of the Presidential Task Force*. Washington: American Enterprise Institute.

———— 1979. *The Regulated Industries and the Economy*. New York: W. W. Norton.

Marx, Karl. 1867. *Capital*. Reprint ed., New York: Modern Library, 1906.

McGraw-Hill. Various years. *Annual Survey of Investment in Employee Safety and Health*. New York: McGraw-Hill.

Mendeloff, John. 1979. *Regulating Safety: An Economic and Political Analysis of Occupational Safety and Health Policy*. Cambridge, Mass.: MIT Press.

Michigan, University of, Institute for Social Research. 1971. *Survey of Working Conditions.* Ann Arbor: University of Michigan, Social Science Archive.

———— 1976, 1977. *Panel Study of Income Dynamics.* Ann Arbor: University of Michigan, Social Science Archive.

Miller, James C. 1976. Statement on behalf of the Council on Wage and Price Stability before OSHA, "Exposure to Inorganic Arsenic."

Miller, James C., and Bruce Yandle. 1979. *Benefit-Cost Analyses of Social Regulation.* Washington: American Enterprise Institute.

Mishan, E. J. 1971. Evaluation of Life and Limb: A Theoretical Approach. *Journal of Political Economy* 79:706–738.

Morrall, John F. 1976a. Statement on behalf of the Council on Wage and Price Stability before OSHA, "Exposure to Coke Oven Emissions."

———— 1976b. Statement on behalf of the Council of Wage and Price Stability before OSHA, "Occupational Noise Exposure."

———— 1979. An Economic Analysis of OSHA's Cotton Dust Standard. Paper presented at 1979 Brookings Institution Conference on the Scientific Basis for Health and Safety Regulations.

———— Forthcoming. *OSHA after Ten Years.* Washington: American Enterprise Institute.

National Commission on State Workmen's Compensation Laws. 1973. *Compendium on Workmen's Compensation.* Washington: U.S. Government Printing Office.

National Safety Council. 1972. *Supervisors' Safety Manual.* Chicago: National Safety Council.

———— 1979, 1981. *Accident Facts.* Chicago: National Safety Council.

Oi, Walter. 1973. An Essay on Workmen's Compensation and Industrial Safety, in *Supplemental Studies for the National Commission on State Workmen's Compensation Laws.* Washington: U.S. Government Printing Office.

———— 1974. On the Economics of Industrial Safety, *Law and Contemporary Problems* 38:538–555.

———— 1975. On Evaluating the Effectiveness of the OSHA Inspection Program. Unpublished manuscript, University of Rochester.

Page, Joseph, and Mary-Win O'Brien. 1973. *Bitter Wages.* New York: Grossman.

Peltzman, Sam. 1975. The Effects of Automobile Safety Regulation, *Journal of Political Economy* 83:677–725.

Raiffa, Howard. 1968. *Decision Analysis: Introductory Lectures on Choices under Uncertainty*. Reading, Mass.: Addison-Wesley.

Raiffa, Howard, and Robert Schlaifer. 1961. *Applied Statistical Decision Theory*. Cambridge, Mass.: MIT Press.

Rice, Dorothy, and Barbara Cooper. 1967. The Economic Value of Life, *American Journal of Public Health* 57:1954-1966.

Schelling, Thomas. 1968. The Life You Save May Be Your Own, in *Problems in Public Expenditure Analysis*, ed. S. Chase. Washington: Brookings Institution.

Scott, Rachel. 1974. *Muscle and Blood*. New York: E. P. Dutton.

Smith, Adam. 1776. *The Wealth of Nations*. Reprint ed., New York: Modern Library, 1937.

Smith, Robert S. 1976. *The Occupational Safety and Health Act: Its Goals and Achievements*. Washington: American Enterprise Institute.

——— 1979a. Compensating Wage Differentials and Public Policy: A Review. *Industrial and Labor Relations Review* 32:339-352.

——— 1979b. The Impact of OSHA Inspections on Manufacturing Injury Rates. *Journal of Human Resources* 14:145-170.

Spence, Michael. 1977. Consumer Misperceptions, Product Failure and Producer Liability. *Review of Economic Studies* 44:561-572.

Stellman, James, and Susan Daum. 1973. *Work Is Dangerous to Your Health*. New York: Vintage Books.

Stokey, Edith, and Richard Zeckhauser. 1978. *A Primer for Policy Analysis*. New York: W. W. Norton.

Thaler, Richard, and Sherwin Rosen. 1976. The Value of Saving a Life: Evidence from the Labor Market. In *Household Production and Consumption*, ed. N. Terleckyz. NBER Studies in Income and Wealth no. 40. New York: Columbia University Press.

Thurow, Lester. 1980. *The Zero-Sum Society*. New York: Basic Books.

Truffaut, Francois. 1967. *Hitchcock*. New York: Simon and Schuster.

U.S. Bureau of Labor Statistics. 1972a. *Handbook of Labor Statistics*, Bulletin 1735. Washington: U.S. Government Printing Office.

——— 1972b. *Work-Injury Rates by Industry*, Bulletin 2097. Washington: U.S. Department of Labor.

——— 1976. *Major Collective Bargaining Agreements: Safety and Health Provisions*, Bulletin 1425-16. Washington: U.S. Department of Labor.

——— 1979. *Work-Related Deaths for 1978*. Washington: U.S. Department of Labor.

———— 1980. *Occupational Injuries and Illnesses in 1978: Summary,* Report 586. Washington: U.S. Department of Labor.

———— 1981. *Occupational Injuries and Illnesses in 1979: Summary,* Bulletin 2097. Washington: U.S. Department of Labor.

———— Various years. *Handbook of Labor Statistics.* Washington: U.S. Government Printing Office.

———— Various years. *Occupational Injuries and Illnesses in the United States by Industry.* Washington: U.S. Government Printing Office.

U.S. Chamber of Commerce. 1980. *Analysis of Workers' Compensation Laws.* Washington: U.S. Chamber of Commerce.

U.S. Council on Wage and Price Stability. 1976. *Comments on the Proposed OSHA Occupational Noise Exposure Standard.* Washington: Council on Wage and Price Stability.

———— 1977a. *Comments on the Proposed OSHA Cotton Dust Standard,* Washington: Council on Wage and Price Stability.

———— 1977b. *Comments on the Proposed OSHA Deep Sea Divers Standard.* Washington: Council on Wage and Price Stability.

———— 1977c. *Comments on the Proposed OSHA Lead Standard.* Washington: Council on Wage and Price Stability.

———— 1978a. *Regulatory Analysis Review Group Report on the Proposed OSHA Acrylonitrile Standard.* Washington: Council on Wage and Price Stability.

———— 1978b. *Regulatory Analysis Review Group Report on the Proposed OSHA Standard for the Identification, Classification, and Regulation of Toxic Substances Posing a Potential Occupational Carcinogenic Risk.* Washington: Council on Wage and Price Stability.

U.S. Department of Commerce. 1980. *Survey of Current Business,* various issues.

U.S. Department of Labor, Occupational Safety and Health Administration. 1979a. *Field Operations Manual,* vol. 5. Washington: U.S. Department of Labor.

———— 1979b. *General Industry Standards.* Washington: U.S. Government Printing Office.

———— 1980a. Identification, Classification, and Regulation of Potential Occupational Carcinogens. *Federal Register* 45:5001–5296.

———— 1980b. *Industrial Hygiene Field Operations Manual,* vol. 6. Washington: U.S. Department of Labor.

———— 1980c. List of Substances Which May Be Candidates for Further Scientific Review and Possible Identification, Classification, and Regulation as Potential Occupational Carcinogens. *Federal Register* 45:53672–53679.

—— 1980d. *Selected Occupational Fatalities Related to Oil/Gas Well Drilling Rigs as Found in Reports of OSHA Fatality/Catastrophe Investigations*. Washington: U.S. Department of Labor.

—— 1980e. *Status: Safety and Health Regulations*. Washington: U.S. Department of Labor.

—— 1981. *Draft Regulatory Analysis and Environmental Impact Statement for the Hazards Identification Standard*. Washington: U.S. Department of Labor.

—— Various years. *The President's Report on Occupational Safety and Health*. Washington: U.S. Government Printing Office.

U.S. Department of Labor, Office of the Assistant Secretary for Policy, Evaluation, and Research. 1980. *Compliance with Standards, Abatement of Violations, and Effectiveness of OSHA Safety Inspections*, Technical Analysis Paper no. 62. Washington: U.S. Department of Labor.

U.S. Office of Management and Budget. 1980. *Budget of the United States Government, Fiscal Year 1981*. Washington: U.S. Government Printing Office.

Viscusi, W. Kip. 1978a. Labor Market Valuations of Life and Limb: Empirical Estimates and Policy Implications. *Public Policy* 26:359–386.

—— 1978b. Wealth Effects and Earnings Premiums for Job Hazards. *Review of Economics and Statistics* 60: 408–418.

—— 1979a. *Employment Hazards: An Investigation of Market Performance*. Cambridge, Mass.: Harvard University Press.

—— 1979b. The Impact of Occupational Safety and Health Regulation. *Bell Journal of Economics* 10:117–140.

—— 1979c. Job Hazards and Worker Quit Rates: An Analysis of Adaptive Worker Behavior. *International Economic Review* 20:29–58.

—— 1980a. Imperfect Job Risk Information and Optimal Workmen's Compensation Benefits. *Journal of Public Economics* 14:319–337.

—— 1980b. Self-Selection, Learning-Induced Quits, and the Optimal Wage Structure. *International Economic Review* 21:529–546.

—— 1980c. A Theory of Job Shopping: A Bayesian Perspective. *Quarterly Journal of Economics* 94:609–614.

——— 1980d. Unions, Labor Market Structure, and the Welfare Implications of the Quality of Work. *Journal of Labor Research* 1:175–192.

——— 1981. Occupational Safety and Health Regulation: Its Impact and Policy Alternatives. In *Research in Public Policy Analysis and Management,* vol. 2, ed. J. Crecine. Greenwich, Conn.: JAI Press, pp. 281–289.

Viscusi, W. Kip, and Richard Zeckhauser. 1979. Optimal Standards with Incomplete Enforcement. *Public Policy* 26:437–456.

Zeckhauser, Richard. 1975. Procedures for Valuing Lives. *Public Policy* 23:419–464.

Zeckhauser, Richard, and Albert Nichols. 1979. The Occupational Safety and Health Administration: An Overview. In *The Study of Federal Regulation of the Senate Committee on Governmental Affairs,* vol. 6. Washington: U.S. Government Printing Office, pp. 163–248.

Zeckhauser, Richard, and Donald Shepherd. 1976. Where Now for Saving Lives? *Law and Contemporary Problems* 40:5–45.

INDEX